Letts

CAMBRIDGE IGCSE® PHYSICS

Revision Guide

Malcolm Bradley

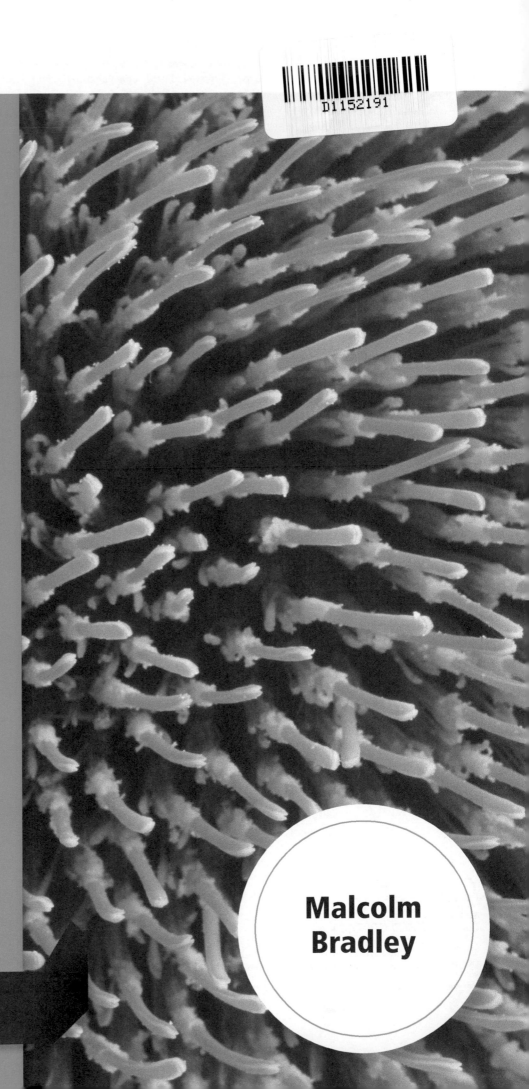

ACKNOWLEDGEMENTS

Cover photo © Photoshot License Ltd / Alamy Stock Photo
Illustrations by Green Hill Wood Studios, Banbury, UK on pages 13, 24, 25, 38, 39, 52, 53, 63, 67, 75, 76
Original illustrations on other pages by Jouve India Private Ltd, and Anne Paganuzzi

Published by Letts Educational
An imprint of HarperCollins*Publishers*
The News Building
1 London Bridge Street
London
SE1 9GF

HarperCollins *Publishers*
1st Floor
Watermarque Building
Ringsend Road
Dublin 4
Ireland

ISBN 978-0-00-821033-5

First published 2017

10 9 8 7 6 5

© HarperCollins*Publishers* Limited 2017

Commissioned by Katherine Wilkinson
Project managed by Sheena Shanks
Edited by Jill Laidlaw and Louise Robb
Proofread by Jess White
Cover design by Paul Oates
Typesetting by Greenhill Wood Studios, Banbury, UK
Production by Natalia Rebow and Lyndsey Rogers
Printed and Bound in the UK using 100% Renewable Electricity at CPI Group (UK) Ltd

MIX
Paper from
responsible sources
FSC www.fsc.org   **FSC™ C007454**

This book is produced from independently certified
FSC™ paper to ensure responsible forest management.

For more information visit: www.harpercollins.co.uk/green

Contents

Chapter 5 Atomic physics

Introduction

Welcome to this revision guide. It will help you to prepare for your Cambridge IGCSE® Physics examinations. It contains all the topics required for the syllabus, but it does not include all the depth of explanation that you will find in your own notes from the course, so it is important to use both when you are revising. Similarly, it is important to use this guide over an extended period of time to ensure that your knowledge is regularly consolidated – it is simply not possible to revise the entire Physics course in a couple of days!

The guide is divided into five sections: General physics, Thermal physics, Properties of waves, including light and sound, Electricity and magnetism, and Atomic physics. The main text describes and explains the content from the syllabus, with Supplement material clearly identified. Do also make sure to study the diagrams – key facts are included within them.

Each section is divided into several smaller parts to help you plan your revision, and each of these is followed by a Quick test so that you can check you have remembered the key points. When you are confident that you have mastered a section, try the Exam-style practice questions which follow – they reflect the type and style of the questions you are likely to get in the exam. Remember that you will usually need to provide a separate point in your answer for each mark. You will also find Revision tips throughout the guide – they offer useful reminders and highlight common errors.

Revision is often more successful if you do something active rather than simply reading. You could try rewriting some of the material in a different form; for example, you could convert a paragraph of text into a series of bullet points or create a set of revision cards, or perhaps you could try to remember the key steps in an experiment by drawing a series of pictures in order. There are many active ways to revise – try some different ones to find out what works well for you.

Wishing you the very best of luck with your Physics revision and examinations.

Length and time

Measuring

Physics is a measuring science. To describe the motion of objects we need reliable measurements of a number of properties. Two common measurements are of the **length** of objects and the **time** it takes for events to happen. We measure lengths with a **rule** and times with a **clock** or other device (such as using **light gates** connected to a **computer** or **data logger**).

Supplement

A micrometer screw gauge measures small lengths accurately, usually measuring to the nearest 0.01 mm. To make a measurement, tighten the micrometer screw gauge around the object – a ratchet reduces the chance of tightening too much. To find the value, the main scale reads to the nearest half a millimetre and then the number of hundredths to be added can be read.

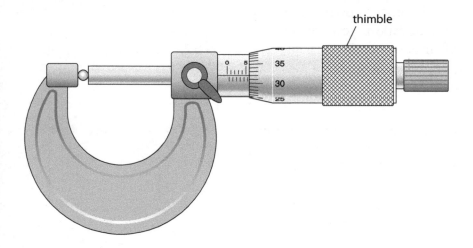

thimble

Accurate timing

Timing short events with a hand-held device such as a stopclock can lead to inaccurate measurements because of the reaction time in starting and stopping the device. To improve the measurements, time the event over multiples, for example time a pendulum swinging for 20 swings, and then divide by the number of events. This gives an average result where the effect of the timing error is much smaller.

Quick test

1. Name the instrument you could use to measure the length of a book.
2. Explain why using a hand-held stopwatch to measure short intervals of time can lead to inaccurate measurements.
3. Describe how to find the thickness of a piece of paper using a rule marked in mm and cm.

Supplement

4. Explain how using a micrometer screw gauge improves the measurement of short lengths.

Graphs of motion

The motion of objects is often described using graphs.

Distance–time graphs

Distance–time graphs show **how far** an object has moved at different **times**.

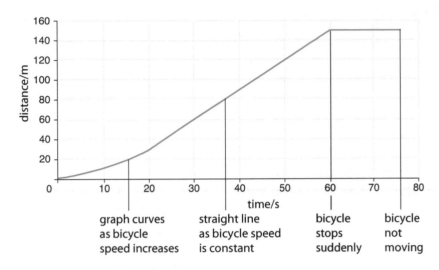

On a distance–time graph:

- a **horizontal straight line** shows the object is **at rest** (not moving)
- a **straight line** (but not horizontal) shows the object is moving at a **constant speed**
- a **line becoming steeper** shows the **speed is increasing**
- a **line becoming less steep** shows the **speed is decreasing**

> **Supplement**
>
> - calculating the **gradient** of a distance–time graph gives the **speed** of the object.

Definition

Speed measures how quickly an object is moving.

$$v = s\,/\,t$$

where:

- v is the speed in metres per second, m/s
- s is the distance travelled in metres, m
- t is the time taken in seconds, s.

Cover speed to find that speed = distance/time

Supplement

Scalars and vectors

The **velocity** of an object is the speed that an object has **and the direction it is travelling in**. For example, a car travelling from London to Brighton could have a speed of 30 m/s but a velocity of 30 m/s **southwards**.

Velocity is an example of a vector quantity. Vector quantities have a **magnitude** (a value) and a **direction**. Other examples include **force**, acceleration and **momentum**.

Speed is an example of a scalar quantity. Scalar quantities have a **magnitude** only. Other examples include mass and **temperature**.

To combine vector quantities, such as combining velocities, draw arrows to scale to represent each one, joining them 'head to tail'. The combined value and direction is given by a single arrow drawn from the 'tail' of the first arrow to the 'head' of the final one.

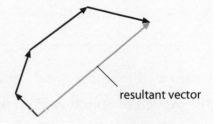
resultant vector

Speed–time graphs

Speed–time graphs display a lot of information.

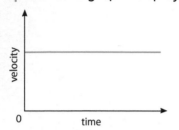 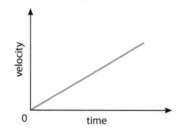 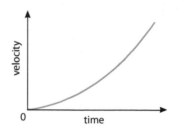

- Calculating the **area under the graph** gives the **distance travelled**.
- The **gradient** of the graph gives the **acceleration** – the steeper the line, the greater the acceleration.

Supplement

- Calculating the **gradient** of a speed–time graph gives the **acceleration** of the object.
- A **curved** line indicates that the acceleration is **changing**.

Definition
Acceleration measures how quickly a velocity is changing. $a = (v - u) / t$ where: - a is the acceleration in metres per second squared, m/s^2 - v is the final velocity in metres per second, m/s - u is the initial velocity in metres per second, m/s - t is the time taken in seconds, s. A **negative acceleration** indicates that the object is **decelerating**.

Revision tip

Make sure you look carefully to see which type of graph you are dealing with.

Quick test

1. On a distance–time graph, what does a horizontal line represent?
2. On a distance–time graph, what does a straight, but not horizontal, line represent?
3. How could you tell from a distance–time graph that the speed of an object was increasing?
4. On a speed–time graph, what does a horizontal line represent?
5. On a speed–time graph, what does a straight, but not horizontal, line represent?
6. Describe how to find the distance travelled from a speed–time graph.

Supplement

7. Describe the difference between a scalar quantity and a vector quantity.
8. List **three** scalar quantities and **three** vector quantities.

Weight is a force, it is measured in **newtons, N**. Weight is caused by a gravitational field, such as the one around the Earth, acting on any object that has **mass**.

Mass is a measure of how much material an object contains. It is measured in **kilograms, kg**.

Weight and mass are linked using this equation:

$W = m \times g$

where:

- W is the weight in newtons, N
- m is the mass in kilograms, kg
- g is the gravitational field strength in newtons per kilogram, N/kg.

> **Revision tip**
>
> Weights and masses can be compared using a balance.

Supplement

As well as measuring how much material an object contains, mass is also a measure of how difficult it is to change the motion of an object. Imagine pushing a young child on a swing at a playground and then pushing a full-sized adult on the same swing – it is more difficult to set the adult going as they have more mass.

Falling objects

The force of gravity causes objects to fall, such as a skydiver or drops of rain.

Near to the Earth, if there was no air resistance, then objects would continue to accelerate as they fall, increasing their speed by a constant 10 m/s each second.

However, in reality there is air resistance and this causes an upward force, which works against the weight of the object. As the speed increases, so does the air resistance. This means that the resultant force downwards is reduced and the object will speed up at a slower rate. Once the air resistance has the same value as the weight of the object the resultant force will be zero and the object continues to fall at a steady speed – this speed is called terminal velocity.

Quick test

1. Explain the difference between weight and mass.
2. Write down the equation linking weight and mass.

Supplement

3. Explain how air resistance causes a falling object to reach a terminal velocity.

Density and pressure

Density

> ### Definition
>
> Density is a measure of how compact a material or an object is.
>
> $\rho = m / V$
>
> where:
>
> - ρ is the density in kilograms per metres cubed, kg/m³
> - m is the mass in kilograms, kg
> - V is the volume in metres cubed, m³.

Measuring density

To find the density of an object or material, two measurements need to be made.

The **mass** of the object is measured using a **balance** or **scales**.

The method for finding the **volume** depends on the **shape** of the object. If the object has a **regular** shape, such as a rectangular block or cylinder, then the volume can be calculated from measurements of the dimensions. For a rectangular block, the volume is given by the width × the length × the height. For a cylinder it is the length of the cylinder × the cross-sectional area of the cylinder (a circle). For an object with an **irregular** shape the volume is found by displacing water. To do this, submerge the object under water and then measure the volume of water displaced (pushed out of the way) using a **measuring cylinder**. If the object is small enough it will fit into the measuring cylinder and the volume of the object is the difference in the volumes recorded on the measuring cylinder scale. The object can also be submerged in a **displacement can** and the displaced water is collected in a measuring cylinder to give the volume of the object. Once you have the mass and volume measurements, calculate the density using density = mass / volume.

Floating and sinking

An object will float if it is less dense than the fluid (liquid or gas) in which it is placed. For example, helium is less dense than air, so a balloon filled with helium floats in the air. However, if an object is more dense than the fluid it is in, it will sink. For example, gold is more dense than water, so a gold ring sinks in a bowl of water.

Pressure

Pressure measures how 'spread out' a force is over an area.

Definition

$p = F / A$

where:

- p is the pressure in pascals, Pa
- F is the force in newtons, N
- A is the area in metres squared, m^2.

So to create a high pressure, you need a large force, a small area, or both. For example, a drawing pin has a small point to 'concentrate' the force applied when you push it in. A low pressure is created in the reverse manner – for example, tractor tyres have a large area to reduce the pressure it exerts on the ground and stop it sinking in.

Measuring pressure

The particles of the air constantly collide with surfaces and this leads to the idea of air pressure.

A **mercury barometer** is an instrument for measuring air pressure. In its simplest form it consists of a tube filled with mercury which is forced up the tube by the air pushing down on the mercury at the base. The higher the air pressure, the taller the column of mercury.

A **manometer** is another instrument for measuring pressure. The U-tube is filled with liquid and the height of the liquid on either side gives a comparison of the pressure on either side.

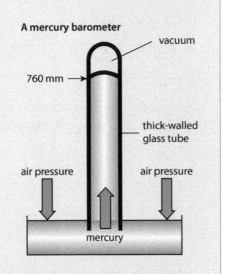

A mercury barometer

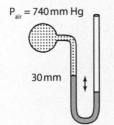

$P_{air} = 740 \, mm \, Hg$

30 mm

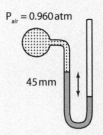

$P_{air} = 0.960 \, atm$

45 mm

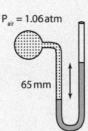

$P_{air} = 1.06 \, atm$

65 mm

The pressure measured by each of these manometers is different, which is indicated by the difference in height of the liquid between the two tubes shown in each one.

The pressure below the surface of a liquid, such as the sea, increases as you go deeper as the weight of the fluid above pushes down on you.

To calculate the pressure at any depth, use the equation:

$$p = h \times \rho \times g$$

where:

- p is pressure
- h is depth
- ρ is density of the liquid
- g is gravitational field strength.

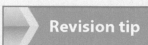

Revision tip

Be careful not to confuse the letter 'p' (for pressure) with the Greek letter ρ, which we use for 'density'.

Quick test

1. Describe how to measure the volume of a regularly shaped object.
2. Describe how to measure the volume of an irregularly shaped object.
3. How do you calculate the density of an object if you know its mass and volume?
4. Write down the equation linking pressure, force and area.
5. Give an example of a force being spread over a larger area to reduce the pressure.
6. Give an example of a force being applied over a smaller area to increase the pressure.

Supplement

7. Write down the equation linking the pressure in a liquid, the depth in the liquid, the density of the liquid and the gravitational field strength, g.

Effects of forces

Forces can change the **shape** or **size** of an object. They can also change the **motion** of an object – causing it to **speed up** or **slow down** or **change direction**.

A resultant force is the **combined** effect of two or more forces acting together.

If the forces are in the **same** direction, then the resultant force is their sum. For example, forces of 20 N and 15 N pulling in the same direction give a resultant force of 35 N.

If the forces are in **opposite** directions, then the forces subtract. For example, forces of 20 N and 15 N pulling in opposite directions give a resultant force of 5 N in the direction of the 20 N force.

If the resultant force on an object is **zero**, then the object will remain **at rest** or, if it is already moving, it will continue at a **constant speed** in a **straight line**.

Supplement

If the resultant force on an object is **not zero,** then the object will **accelerate in the direction of the resultant force**. This will cause the object to increase or decrease its speed if the resultant force is in line with the object's motion, or will cause the object to travel in a **circle** if the resultant force is at right angles to the motion.

The acceleration caused by a resultant force is calculated from:

$F = m \times a$

where:

- *F* is the resultant force
- *m* is the mass of the object
- *a* is the acceleration.

> **Revision tip**
>
> The units of acceleration are m/s² (metres per second squared). Make sure you memorise this..

Friction is a force between surfaces which acts against any motion between the surfaces. Friction causes surfaces to become hot. Air resistance is also a form of friction.

> **Revision tip**
>
> Be careful **not** to say 'energy lost as friction' – friction is a **force**, not a form of energy.

Experiment

Extension–load graphs

To gather the data to plot an extension–load graph (for example for a spring or a rubber band) follow these steps.

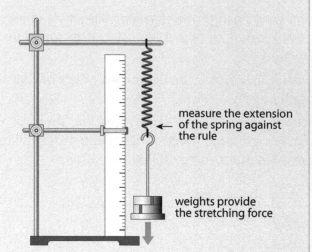

measure the extension of the spring against the rule

weights provide the stretching force

1. Hang the spring (or other material) from a clamp stand.
2. Measure its length with a rule held vertically next to it.
3. Add a 100 g mass from the spring.
4. Measure the new length.
5. Calculate the **extension** of the spring by subtracting the **original** length.
6. Repeat this process for additional 100 g masses.
7. Calculate the **load** provided by each mass using $W = m \times g$ (each 100 g mass provides a load of 1 N).
8. Plot a graph of **extension** (*y*-axis) against **load** (*x*-axis).

Hooke's law states that the extension is proportional to the load. This is shown by the straight line part of the graph.

For the Hooke's law part of the graph:

$$F = k \times x$$

where:

- F is the load
- x is the extension
- k is the **spring constant**, a number which is calculated from the gradient of the graph.

The graph also shows a **limit of proportionality** – this is where the graph starts to curve and Hooke's law no longer applies.

Quick test

1. List **three** effects that forces can have on an object.
2. What is a resultant force?
3. Describe how to combine two forces to find the resultant force.
4. Describe the motion of an object if the resultant force is zero.
5. Write down the units of acceleration.
6. Describe how to investigate the extension of a spring with different loads.

Supplement

7. Describe the motion of an object if the resultant force is not zero.
8. State Hooke's law.

Turning effect

Definition

The moment of a force is a measure of the force's **turning effect**.

Moment = $F \times d$

where:

- moment is measured in newton metres, Nm
- F is the force in newtons, N
- d is the perpendicular distance from the pivot in metres, m.

Simple examples of moments would be opening a door or using a claw hammer to remove a nail. To increase the turning effect, the force can be increased or the distance from the pivot increased. Imagine trying to open a door pushing very close to the hinges – the distance from the pivot (the hinges) is small so a large force is needed. Pushing at the handle – well away from the hinges – means a smaller force is needed for the same turning effect.

The principle of moments says that, if an object is **balanced**, then the **sum** of the **clockwise moments** must equal the **sum** of the **anticlockwise moments**.

Worked example

Phil and Ben are sitting on a seesaw. The seesaw is balanced on a pivot. Calculate Phil's weight.

Ben is causing a clockwise moment of 400 N × 3 m. Phil is causing an anticlockwise moment of W × 2 m.

The see-saw is balanced, so, taking moments about the pivot:

the sum of the clockwise moments = the sum of the anticlockwise moments

400 × 3 = W × 2

W (Phil's weight) = 600 N.

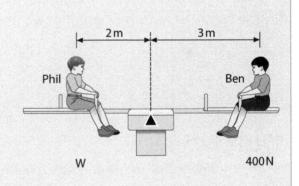

Definition

An object is in equilibrium if the resultant force on it is **zero AND** there is no resultant turning effect on it (the resultant **moment** is zero).

Centre of mass

The **centre of mass** of an object is the point where we consider all the weight of the object to be.

This helps to describe the **stability** of an object. If the object is tipped so that the centre of mass moves outside the base of the object, then it will topple over when it is released – it will be **unstable**. However, if the centre of mass remains inside the base when it is tipped, then it will return to its original position when it is released.

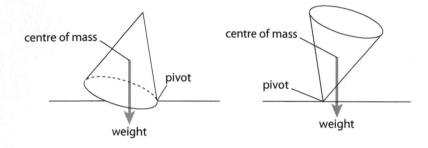

Experiment

To find the centre of mass of a **plane lamina** (such as a piece of card) follow these steps.

1. Hang up the card.
2. Suspend a mass from the same place.
3. Mark the position of the thread – the centre of mass is somewhere along this line.
4. Repeat steps 1 to 3 with the card suspended from a different point.
5. The centre of mass is where the two lines meet.

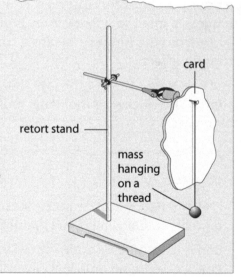

Quick test

1. Write down the equation linking moment, force and perpendicular distance from the pivot.
2. State two ways to increase the moment of a force.
3. How does the principle of moments link clockwise moments and anticlockwise moments?
4. State the two conditions necessary for an object to be in equilibrium.
5. Explain what the phrase 'centre of mass' means.
6. Describe how to find the centre of mass of a piece of card.

Supplement

Momentum

Momentum is the term we use for what Sir Isaac Newton called 'quantity of motion'. It combines the mass and velocity of an object.

Definition

$p = m \times v$

where:

- p is the momentum in kilogram metres per second, kg m/s
- m is the mass in kilograms, kg
- v is the velocity in metres per second, m/s.

Impulse

Impulse describes the **change** in velocity, $mv - mu$, where m = mass, u = starting velocity and v = final velocity.

Impulse can also be calculated from $F \times t$, where F = force acting and t = time for which the force acts.

These two can be combined into $Ft = mv - mu$.

Revision tip

This idea is very helpful in situations such as car safety. In a collision, if the momentum change is spread over a longer time (making t larger) then the forces involved will be much smaller (resulting in less severe injuries). This is the principle behind features such as seat belts, air bags and crumple zones.

The law of conservation of momentum

The **law of conservation of momentum** states that the total momentum in a closed system is constant. This allows us to calculate velocities before and after an event.

Worked example

Two cars collide as shown in the diagram. After the collision both cars are stationary.

Calculate the velocity of the 750 kg car before the collision.

Before · v m/s 750 kg · 15 m/s 500 kg · After

Let v = velocity of the 750 kg car before the collision.

Total momentum before collision = total momentum after collision

$(750 \times v) + (500 \times -15) = (750 \times 0) + (500 \times 0)$

Note that velocity is a vector quantity, so −15 m/s is 15 m/s **towards the left**.

$(750 \times v) - 7500 = 0$

$750 \times v = 7500$

$v = 10$ m/s **to the right** (since it is a positive velocity)

Quick test

Supplement

1. Write down the equation linking momentum, mass and velocity.
2. Momentum is a vector quantity. Explain what this means.
3. Calculate the momentum of a racing car of mass 600 kg travelling at 75 m/s.
4. State the law of conservation of momentum.
5. Explain why a gun recoils backwards if it fires a bullet forwards.

Energy, work and power

Energy

Energy can be transferred in different ways and stored in different forms.

Form of energy	Description	Examples
kinetic	energy of motion, all moving objects possess **kinetic energy**	• a moving car • particles in a gas
gravitational potential	energy stored because of the position of an object – lifting an object higher increases its gravitational **potential energy**	• throwing a ball in the air • walking up stairs
chemical	energy stored in chemicals that can be released in a chemical reaction	• an electric battery • energy stored in food
elastic (strain)	energy stored in an object because it is under strain – it has been stretched	• a stretched spring • a rubber band
nuclear	this is energy stored in the nucleus at the centre of an atom	• nuclear reactors in power stations
internal	this is the energy contained within an object that makes it hot or cold – it is the energy of the vibrations and motion of the particles	• the heat energy in a hot cup of coffee

Energy can be transferred in different ways.

Method of energy transfer	Example
by forces	• throwing a ball
by electric circuits	• lighting a lamp • running a motor
heating	• boiling water over a fire
by waves	• sending radio signals • cooking food in a microwave oven

Although energy can appear in different forms in different situations, the **principle of conservation of energy** states that the **total** energy in a system always remains the same. This means that if one part of a system is losing energy, then some other part must be gaining energy.

Supplement

Although the principle of conservation of energy states that the total energy remains constant, the energy tends to become more 'spread out' among different objects as a result of energy transfer processes that take place. This will leave the energy more **dissipated** and less useful as it is less concentrated.

Definitions

Kinetic energy and gravitational potential energy are calculated using the following equations:

$$\text{kinetic energy} = \tfrac{1}{2}m \times v^2$$

where:

- m is the mass of the object in kilograms, kg
- v is the velocity of the object in metres per second, m/s

$$\text{gravitational potential energy} = m \times g \times \Delta h$$

where:

- m is the mass of the object in kilograms, kg
- g is the gravitational field strength in newtons per kilogram, N/kg
- Δh is the change in the height of the object in metres, m.

Work

Work is done when a force is used to move an object. For example, if I push a book along a table, I lose some chemical energy (from my food) but the book gains kinetic energy. This is described as **doing work** on the book.

The work done in any situation is always equal to the energy transferred.

The amount of work done (energy transferred) increases if the size of the force increases or if the distance over which the force acts increases.

Definition

Work is calculated from:

$$W = F \times d = \Delta E$$

where:

- W is the work done in joules, J
- F is the force applied in newtons, N
- d is the distance moved in the direction of the force in metres, m
- ΔE is the energy transferred in J.

Power

Power measures how **quickly** energy is transferred. The more energy that is transferred per second, the higher the power.

Supplement

Definition

Power is calculated from:

$$P = \Delta E\, /\, t$$

where:

- P is the power in watts, W
- ΔE is the energy transferred in joules, J
- t is the time taken for the energy transfer in seconds, s.

Worked example

An 'energy-efficient' light bulb has a power rating of 11 W. Calculate the energy transferred by the bulb in two minutes.

Energy transferred = power × time

= 11 × 120 seconds (remember to change the time from minutes into seconds!)

= 1320 J

Quick test

1. Describe examples of energy transferred by **(a)** forces, **(b)** electric circuits, **(c)** heating, **(d)** waves.
2. State the law of conservation of energy.
3. When is work done?
4. How does the amount of work done compare to the amount of energy transferred?

Supplement

5. Explain what the word dissipated means in connection with energy transfer.
6. Write down the equations for calculating kinetic energy and gravitational potential energy.
7. Write down the equation linking work done, force applied and distance moved.
8. Write down the equation linking power, energy transferred and time taken for the transfer.

Energy resources

Energy resources provide a source of energy that we can usefully use to provide energy for society. Commonly we use these energy resources to generate **electricity**. Electrical energy is a particularly useful form of energy in everyday life since it is relatively easy to distribute, transfers energy very quickly and can be used to provide energy for a wide range of appliances. For details of the process of transferring the energy to electrical energy in a power station, see the section on **generators**.

Renewable energy resources can be replaced in relatively short time scales, whereas **non-renewable** ones cannot (often taking millions of years).

For all energy resources, not all of the available energy is transferred to **useful** energy. Efficiency is a measure of how much energy can be usefully transferred – the higher the efficiency, the more energy is useful.

Supplement

Definition
Efficiency is calculated using

$$\text{efficiency} = \frac{\text{useful energy output}}{\text{energy input}} \times 100\% \text{ and}$$

$$\text{efficiency} = \frac{\text{useful power output}}{\text{energy input}} \times 100\%$$

Revision tip

Remember that the original source of the energy for all these energy resources is the Sun, **except** for geothermal, nuclear and tidal energy.

Energy resource	How it works	Advantages	Disadvantages	Renewable?
fossil fuels (coal, oil, natural gas)	chemical energy is released when the fuel is burned	can provide large amounts of energy to meet demand	pollution, particularly greenhouse gases take millions of years to replace	no
nuclear fission	binding energy of a nucleus is released when large nuclei are split into two	can produce large amounts of energy	waste materials can be highly radioactive	no
biofuels (biomass, biogas, etc.)	these are burned to release chemical energy	can be replaced relatively quickly and can provide a useful way to use animal waste	large amounts of land would need to be given over to growing the crops for fuel, leaving less land available to grow food	yes

tidal	set up across wide estuaries, the tide rising and falling turns generators	reliable resource, free to use once the tidal barrage has been built	can be disruptive to animal and bird habitats\n\nonly limited scale, cannot provide large amounts of energy	yes
wave	waves on the surface of the sea drive a generator	free once the generators have been built	not reliable\n\ncan only produce limited amounts of energy	yes
hydroelectric	water is stored behind a dam and then allowed to run downhill to turn the generators	renewable resource, little pollution	can involve flooding entire valleys to create the reservoir to store the water\n\nneeds hilly environment	yes
geothermal	cold water is pumped into the ground and heated by the Earth to produce steam	free and effectively renewable resource once the infrastructure has been built	only effective in areas where a heat source is accessible underground	effectively yes since the Earth stores large amounts of energy
solar panels (water)	water is pumped through pipes in the panel and heated by the Sun	once installed, hot water is essentially free, reducing the usage of other sources	cannot be used at night and not reliable during the day	yes
solar cells	cells convert the energy in sunlight directly to electricity	once installed, sunlight is free\n\ncan be used to charge batteries	cannot be used at night and not reliable during the day	yes
wind	turbine blades turn electric generators	once installed, the wind is free	cannot be used if the wind is too light or too strong\n\ncan be noisy\n\ncan disrupt bird life	yes

Quick test

1. Explain what the phrase 'energy resource' means.
2. What are energy resources commonly used to generate?
3. Describe the difference between renewable and non-renewable energy resources.
4. List **two** non-renewable energy resources.
5. List **two** renewable energy resources.
6. For each answer in Q4 and Q5, state **one** advantage and **one** disadvantage of each energy resource.

1 A person walks to a local shop and then returns home. The graph shows how their distance from home changes during the walk.

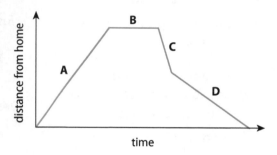

(a) (i) State which part of the graph, **A**, **B**, **C** or **D**, shows them walking the fastest. [1]

(ii) Explain your choice. [1]

(b) The distance from their home to the shop is 600 m and the journey to the shop and back takes 10 minutes.

Calculate the average speed for the whole journey. [3]

Supplement

(c) The average velocity for the whole journey is different to your answer in (b).

(i) Describe the difference between speed and velocity. [2]

(ii) State the average velocity for the whole journey. [1]

(iii) Explain your answer to (ii). [1]

2 The figure shows a Formula 1 racing car.

(a) Describe **two** features of the Formula 1 car that make it stable and unlikely to tip over. [2]

(b) The mass of the car is 700 kg and the area of the tyres in contact with the ground is 0.5 m².

(i) Calculate the weight of the car and give the unit. [2]

(ii) Calculate the pressure in pascals on the ground caused by the car. [2]

Supplement

(c) At the start of a race, the car accelerates from 0 m/s to 46 m/s in 4.0 s.

Calculate the acceleration of the car and give the unit. [3]

3 A skydiver jumps from a plane, free falls initially and then opens a parachute. The graph shows the velocity changes with time during the fall.

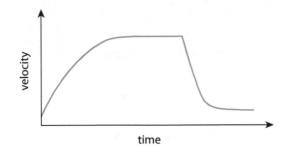

(a) The skydiver has a mass of 70 kg.

    (i) Name the force that causes the skydiver to accelerate downwards.   [1]

    (ii) Calculate the size of this force and give the unit.   [2]

Supplement

(b) The skydiver reaches a terminal velocity.

    (i) On the graph, label this terminal velocity with an *X*.   [1]

    (ii) Explain why the skydiver reaches a terminal velocity.   [4]

(c) (i) On the graph, mark the point where the skydiver opens their parachute with a *Y*.   [1]

    (ii) Explain why the skydiver reaches a new, different terminal velocity.   [3]

4 The table shows some information about two types of light bulb. Both types of bulb have the same brightness.

Type of bulb	Power in W	Lifetime in hours	Cost in $
LED bulb	12	40 000	20.00
filament bulb	60	1000	2.50

(a) Use the information in the table to explain how using LED bulbs can cost less than using filament bulbs.   [3]

Supplement

(b) The efficiency of the LED bulb is 80%.

    (i) Show that the total energy transferred by the LED bulb in 10 minutes is approximately 7000 J.   [2]

    (ii) Calculate the energy transferred by the LED bulb as light in 10 minutes.   [2]

    (iii) Suggest what happens to the rest of the energy.   [1]

5   Electricity is generated using renewable and non-renewable energy sources.

   (a) Name **two** renewable energy sources.   [2]

   (b) Name **two** non-renewable energy sources.   [2]

   (c) Between 1990 and 2005, the percentage of electricity generated in one country using renewable energy sources increased from 2% to 4%.

      (i) Suggest why the use of renewable energy sources increased.   [2]

      (ii) Suggest why the percentage of electricity generated from renewable sources is so small.   [2]

Simple kinetic molecular model of matter

States of matter

The **state** of material describes whether it is a solid, a liquid or a gas.

The table shows the key features of each state.

	Solids	Liquids	Gases
Shape	keep a fixed shape	take the shape of the container	take the shape of the container
Volume	keep a fixed volume	keep a fixed volume	spread out to fill the volume of whatever container the gas is in

Molecular model

The **kinetic** molecular model of matter is the idea that **all matter is made of particles called molecules** which combine and move in particular ways to explain the properties of matter that we observe.

Evidence to support this model comes from Brownian motion – this is where particles in a suspension (for example smoke particles in the air) move in a random way because they are hit by the smaller molecules that we can't see.

Supplement

> **Revision tip**
>
> Larger, more massive particles can be moved around by being hit by light, fast-moving particles.

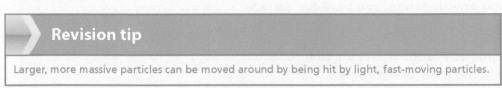

	Solid	Liquid	Gas
Arrangement of molecules	Regular pattern, closely packed together, molecules held in place	Irregular, closely packed together, molecules able to move past each other	Irregular, widely spaced, molecules able to move freely
Diagram			
Motion of molecules	Vibrate in place within the structure	'Slide' over each other in a random motion	Random motion, faster movement than the other states

The molecular model allows us to **explain** the properties of solids, liquids and gases.

	Solids	Liquids	Gases
Shape	keep a fixed shape ... ... because the molecules are locked into position in the lattice by strong attraction forces	take the shape of the container ... ... because the molecules are able to 'slide' over each other as the forces between them are not so strong, so they can fill the spaces into the shape of the container	take the shape of the container ... ... because the molecules are free to move in all directions as the forces between the molecules are so weak
Volume	keep a fixed volume ... ... because the forces between the molecules are strong and hold them in place	keep a fixed volume ... ... because the forces between the molecules are strong enough to hold them close together, although not strong enough to hold them in a fixed place	spread out to fill the volume of whatever container the gas is in ... ... because the forces between the molecules are so small

Evaporation

Definition

How does evaporation happen?

- Molecules in a liquid have a range of energies.
- The molecules with the most energy can escape from the surface.

How does this affect the liquid?

- The average energy of the molecules still in the liquid is reduced.
- So the temperature of the liquid is lower – it cools.

Supplement

What happens to objects in contact with an evaporating liquid?

- Evaporation causes the liquid to cool.
- So energy is transferred into the liquid from objects in contact with it.
- For example, this is how sweating cools your skin.

What factors affect the rate of evaporation?

- Liquids evaporate more quickly if the temperature is higher, they have a larger surface area or if there is a draught.

Pressure changes in gases

If a fixed mass of gas is held in a container at **constant volume**, then increasing the temperature **increases** the pressure. This is because at higher temperatures the molecules move more quickly so they collide with the container walls **more often** and with **greater force**.

Supplement

When the molecules collide with the walls they bounce off – their velocity changes as their direction changes. If the velocity changes, then the momentum changes as well (since $p = m \times v$). This momentum change causes a force on the container walls and pressure is then calculated by dividing the value of this force by the area of the walls.

If a fixed mass of gas is held in a container at **constant temperature**, then increasing the volume **decreases** the pressure. This is because the molecules are moving at the same average speed (constant temperature) so they collide with the container walls **less often**, so the average force on the walls is smaller.

Supplement

For a fixed mass of gas at a constant temperature

pressure × volume = constant

This is an example of an inversely proportional relation – if the volume of gas is doubled, the pressure will be halved, for example.

Quick test

1. Name the three states of matter.
2. Describe how Brownian motion supports the idea that matter is made of particles.

Supplement

3. Use ideas from the molecular model to explain why liquids take the shape of the container.
4. Use ideas from the molecular model to explain why solids have a fixed size.

5. A liquid cools as it evaporates – explain why.

Supplement

6. List three factors that affect the rate of evaporation.

7. Describe how the pressure of a fixed volume of gas changes when its temperature changes.
8. Explain why the pressure in a gas decreases when the volume increases, as long as the temperature is constant.

Thermal properties and temperature

Thermal expansion of solids, liquids and gases

When thermal energy is transferred to materials, they expand (their volume increases). (There are only a few exceptions to this – a notable exception is **water**, which **contracts** as it cools from 100 °C to 4 °C, but then expands as it cools from 4 °C to 0 °C.)

Expansion can be useful – for example, the expansion of the liquid in a thermometer allows us to measure temperature. Or it can cause problems – water expands as it freezes to form ice and this can cause pipes to break.

Supplement

Expansion happens because the molecules gain additional energy. In solid materials, this causes the molecules to vibrate more vigorously in position, but because the forces holding the particles together are large, the expansion of the material overall is quite small. In liquids, the forces are smaller so liquids generally expand at a greater rate than solids. Gases expand the most as the forces holding molecules together are negligible.

Measurement of temperature

Temperature can be measured using any physical property that changes with temperature. For example, it could be the volume of a liquid or the electrical resistance of a wire.

A common example of temperature measurement is to use a liquid-in-glass thermometer. In this a liquid, often alcohol or mercury, expands and rises up a narrow tube as the temperature increases.

The thermometer needs to be calibrated – the numbers on the scale need to be in the correct places. To do this, two **fixed points** are identified – for the Celsius scale these points are at the freezing point (0 °C) and boiling point (100 °C) of water. Once these points are found the rest of the scale can be marked in.

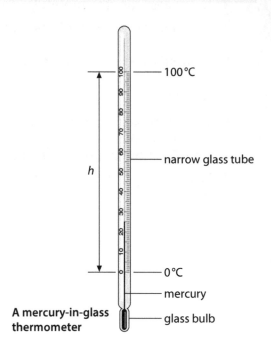

100 °C

narrow glass tube

h

0 °C

mercury

A mercury-in-glass thermometer

glass bulb

Supplement

The liquid-in-glass thermometer has a **thin glass bulb** at the bottom which improves the **sensitivity** of the thermometer (it can respond to smaller changes in temperature). The sensitivity is also improved by making the tube **thin** – this means that smaller changes in volume cause measurable changes on the scale.

Different liquids can be used to give thermometers that will operate over a different **range** of temperatures. For example, mercury cannot be used below –39 °C as it freezes at that point. However, alcohol can be used as low as –114 °C.

It is common for the scale on a thermometer to be divided up evenly. This means it is important that the liquid expands by the same amount for each 1-degree rise in temperature. Liquids that expand in this way have a **linear** response.

A **thermocouple** is an electrical thermometer.

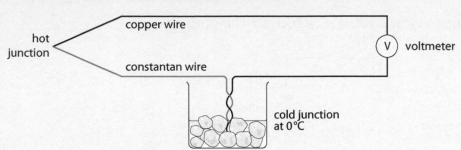

Advantages of thermocouples include:

- they can be made very small and respond quickly to changes in temperature
- they can measure temperatures to above 1000 °C
- the scale to be read can be set away from the thermocouple, allowing for remote measurement.

Thermal capacity (heat capacity)

As thermal energy is added to an object the **internal energy** of the object (the kinetic and potential energy of the molecules) increases.

The thermal capacity of an object is a measure of how much heat energy is required to change the temperature of an object by 1 °C. It increases if the mass of the object increases and it also depends on the material the object is made from.

Supplement

Specific heat capacity (symbol: c) of a material is the energy required to change the temperature of 1 kg of the material by 1 °C.

Energy changes involved in heating are calculated from

$$\Delta E = m \times c \times \Delta T$$

where:

- ΔE is the energy change in joules, J
- m is the mass in kilograms, kg
- c is the specific heat capacity in joules per kilogram degree Celsius, J/kg °C
- ΔT is the temperature change in degrees Celsius, °C.

Thermal capacity is calculated from thermal capacity = $m \times c$, since thermal capacity is always related to a 1-degree change in temperature.

Experiment

Measuring the specific heat capacity of a substance such as aluminium

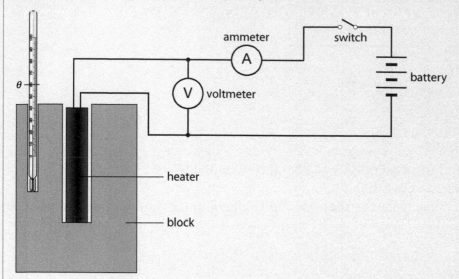

The block is heated for a measured time and the temperature rise is recorded. The readings of the ammeter and voltmeter are also recorded.

The electrical energy supplied by the heater is calculated from $E = V \times I \times t$ (where V is the voltage, I is the current and t is the time the heater was switched on **in seconds**).

The specific heat capacity, c, is then calculated from $c = (V \times I \times t) / (m \times \Delta T)$

Melting and boiling

As a substance gains thermal energy it will change **state** at particular temperatures. The change between solid and liquid happens at the **melting point** and the change between liquid and gas at the boiling point. While the substance changes state there will be an energy transfer into or out of the material, **but the temperature does not change**.

At the melting point, thermal energy added during **melting** breaks up the regular structure of the molecules in a solid to form the more mobile patterns in a liquid. The reverse process, from liquid to solid, is called **freezing** or **solidification**.

At the boiling point, thermal energy added during **boiling** separates the molecules in the liquid structure into the randomly moving separated molecules of a gas. The reverse process, from gas to liquid, is called **condensation**.

 Revision tip

Boiling and evaporating are different. In evaporation, **a few** of the molecules have enough energy to leave the liquid. In boiling, the **average** energy of the molecules is enough to leave – boiling happens at a **specific temperature**, **evaporation** can happen at any temperature.

The thermal energy connected with changes of state is called **latent heat**. The specific latent heat of a substance is the energy transferred to change the state of **1 kg** of the substance without changing the temperature.

For melting/freezing, it is called the specific latent heat of **fusion**.

For boiling/condensing, it is called the specific latent heat of **vaporisation**.

In both situations, the energy transferred is calculated:

$\Delta E = m \times l$

where:

- ΔE is the energy change in joules, J
- m is the mass in kilograms, kg
- l is the specific latent heat in joules per kilogram, J/kg.

Experiment

Measuring specific latent heat

To find the specific latent heat of fusion, use a low voltage heater to melt some ice. The energy transferred can be calculated using $E = V \times I \times t$ (see the box for measuring specific heat capacity) and the mass of ice that melts needs to be measured using a balance. The specific latent heat of fusion is calculated using $l = (V \times I \times t) / m$.

To find the specific latent heat of vaporisation, use the same procedure and equipment, but this time get the water to the boiling point before starting the timing. This time the mass of water that boils away needs to be measured. The calculation is the same as for the specific latent heat of fusion.

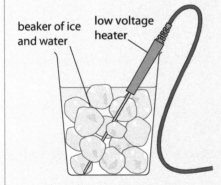

beaker of ice and water   low voltage heater

Quick test

1. What happens to the volume of a material when it expands?
2. Give an example where expansion is useful.

Supplement

3. Use ideas about molecules and energy to explain why solid materials expand when heated.

4. Describe how a thermometer can be calibrated.

Supplement

5. If a liquid used in a thermometer has a *linear* response, what does this mean?
6. List three advantages of thermocouples.

7. What does the thermal capacity of an object measure?

Supplement

8. Describe how to measure the specific heat capacity of a material.

9. Explain why the temperature does not change during melting and boiling.

Supplement

10. Describe the difference between boiling and evaporating.
11. Describe how to measure the specific latent heat of a material.

Thermal processes

Conduction

Heat (thermal energy) can transfer from one particle to the next in a material – this process is called conduction. Heat can be transferred in this way at different speeds. A material where this process happens quickly is called a **conductor**. A material where this process happens quickly is called a **good thermal conductor**. A material where heat is transferred slowly is called a **bad thermal conductor**. Conduction happening at different rates can be shown in this experiment.

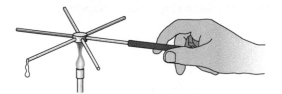

In this experiment to show conduction, the rods are made of different metals, so the heat conducts along them at different rates. The better the conductor, the quicker the wax at the end of the rod melts.

Supplement

Particles in a solid material are constantly vibrating. If heat energy is added to one end of an object the particles vibrate with greater amplitude. Because the particles in a solid are connected by chemical bonds, some of the energy is transferred to neighbouring particles and the energy gradually spreads along the whole object – it gets hotter.

Metals are particularly good conductors of heat because they also have delocalised electrons that can also move through the object, transferring energy as they go.

Convection

Convection is a particularly important method of transferring heat energy through **fluid** materials (liquids and gases). When a fluid is heated, the particles move apart slightly, making the fluid less **dense**. Parts of the fluid that are less dense float upwards, transferring the heat energy. Regions of the fluid that are denser (cooler) sink downwards. This sets up a convection current.

Convection can be demonstrated in this experiment.

Heating potassium permanganate crystals in water to show convection.

Radiation

Heat energy can also be transferred directly by waves of energy – this is **infra-red** radiation, part of the electromagnetic spectrum. Infra-red radiation does not need particles to transfer the energy – this is how heat energy reaches the Earth from the Sun.

All objects emit (give out) and absorb (take in) infra-red radiation, but the rate of energy transfer depends on several factors. Dull, dark colours (such as matt black) make the best absorbers and emitters, while shiny, light colours (such as gloss white) are the worst. However, shiny surfaces make the best **reflectors** of infra-red radiation.

> ### Revision tip
>
> Many students realise that dull, dark surfaces absorb heat radiation the best – think of a black car on a hot summer day – but remember that dull black is also the best **emitter** as well.

Leslie's cube can be used to show how different surfaces affect energy transfer by radiation.

The sides of the cube have surfaces of different colours and textures. As they are all heated by the same hot water, any differences in heat radiated must be due to the surface.

The amount of heat radiated by an object also depends on the **surface temperature** of the object (higher temperatures radiate heat at a higher rate) and the **surface area** of the object (bigger surface areas radiate more heat).

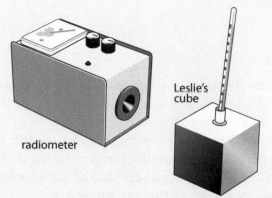

A Leslie's cube. The meter measures the amount of radiation that is emitted by each surface.

Consequences of energy transfer

There are many examples of thermal energy transfer. Our understanding of these processes allows us to design systems to promote the transfer, for example heating a building, or reduce the energy flow, such as insulating a house.

Example 1 – Heating a room

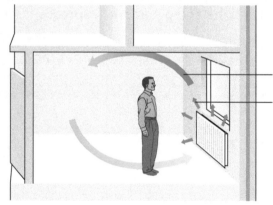

convection current

infra-red radiation

A side view of a room with a hot-water radiator underneath the window. You will see from this that the convection current is far more efficient at heating the top of the room than it is at heating the person standing in front of the radiator.

Example 2 – The vacuum flask

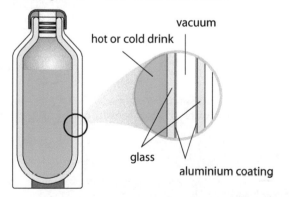

vacuum

hot or cold drink

glass

aluminium coating

The vacuum flask is designed to reduce heat transfer. The **double wall**, separated by a **vacuum**, reduces conduction and convection (since there are no particles) and the **silvered walls** prevent radiation. The stopper reduces heat transfer when the lid is closed.

Quick test

1. For heat transfer, explain the terms 'good thermal conductor' and 'bad thermal conductor'.

Supplement

2. Explain why metals are particularly good conductors.

3. Explain why hot air rises.
4. What is a convection current?
5. Which part of the electromagnetic spectrum is heat radiation?
6. Which types of surfaces absorb and emit heat radiation best?

Supplement

7. Describe how surface temperature and surface area affect the rate that heat is radiated.

8. Describe how the design of a vacuum flask reduces heat transfer.

1 (a) Complete the table to describe the properties of solids, liquids and gases.
One line has been done for you.

	Fixed volume	Fills the container	Fixed shape	Takes the shape of the container
Solids	✓		✓	
Liquids				
Gases				

[2]

Supplement

(b) Use ideas about particles to explain why solid materials have a fixed volume and a fixed shape. [2]

(c) A runner sweats as he exercises. The sweat evaporates from his skin and this helps him keep cool.

(i) Use ideas about particles to explain how a liquid evaporates. [3]

(ii) Explain why sweat evaporating from his skin cools the runner. [3]

2 The diagram shows a potato cooking in a saucepan of hot water.

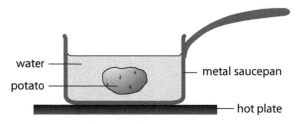

water —
potato —
— metal saucepan
— hot plate

(a) Explain why it is an advantage to make the saucepan from metal. [2]

Supplement

(b) Describe how heat energy is conducted from the hot plate through the metal saucepan. [2]

(c) Describe how heat energy is transferred from the saucepan to the potato by the water. [3]

(d) During cooking, some of the water evaporates and some of the water boils away.
Describe **one** similarity and **one** difference between evaporating and boiling. [2]

3 The diagram shows a liquid-in-glass thermometer.

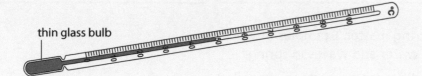

thin glass bulb

(a) State the physical quantity that a thermometer measures. [1]

(b) Explain why the liquid rises up the tube when the thermometer is placed in hot water. [3]

(c) State **two** advantages of the glass bulb having a *thin* wall. [2]

(d) The thermometer has a *linear* scale. Explain what this means. [1]

4 The diagram shows the air movement near a beach during the day.

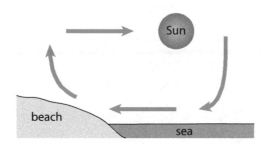

(a) Explain why the air moves as shown in the diagram. [4]

Supplement

(b) The sand and the sea are heated at the same rate by the Sun, but the temperature of the sand is higher than the temperature of the sea. Use ideas about specific heat capacity to explain why. [2]

(c) A student does an experiment to measure the specific heat capacity of sand. She transfers 49 800 J of energy to a 2 kg sample of sand and finds that the temperature rises from 20 °C to 50 °C. Calculate the specific heat capacity of the sand, giving a suitable unit. [4]

General wave properties

Waves transfer energy without transferring matter. Examples of wave motions include light, sound, waves on water and waves in springs.

Waves are formed by vibrations in two different ways; these are called longitudinal and transverse.

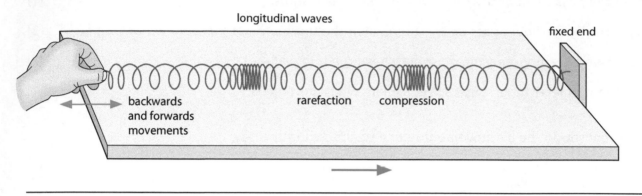

longitudinal waves

fixed end

backwards and forwards movements

rarefaction   compression

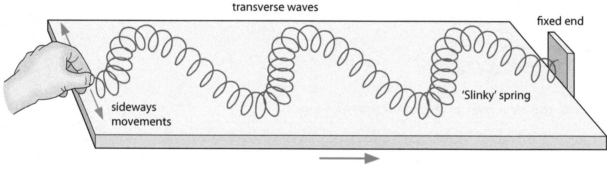

transverse waves

fixed end

sideways movements

'Slinky' spring

direction of wave travel

In longitudinal waves, the vibrations are **parallel** to the energy transfer. Sound waves are examples of longitudinal waves.

In transverse waves the vibrations are **perpendicular** (at right angles) to the energy transfer. Light is an example of a transverse wave.

To describe waves, we use the following key words.

Key word	Unit	Meaning
speed	metres per second, m/s	the distance travelled by the wave in one second
frequency	hertz, Hz	the number of waves passing a point in one second
wavelength	metres, m	the distance from one peak to the next
amplitude	metres, m	the maximum displacement from the equilibrium position

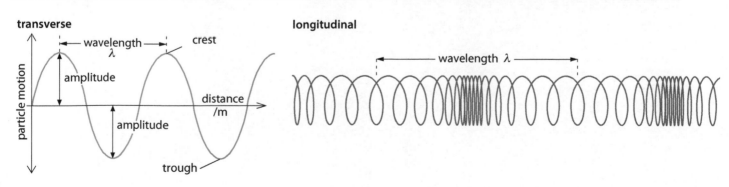

transverse

particle motion

wavelength λ

crest

amplitude

distance /m

amplitude

trough

longitudinal

wavelength λ

In wave diagrams, the single lines showing the direction of the waves are called **rays**. The parallel lines, showing consecutive peaks of the waves, are called **wavefronts**.

As well as transferring energy without transferring matter, all waves share some other properties. These can be shown with water waves using a ripple tank.

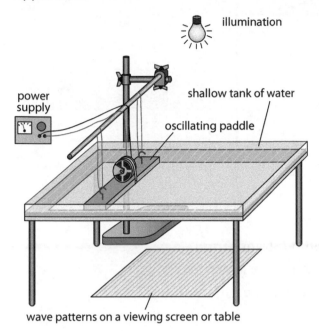

wave patterns on a viewing screen or table

1. All waves **reflect**.

   In reflection, the angle of incidence (labelled *i*) is equal to the angle of reflection (labelled *r*).

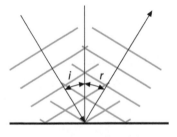

2. All waves refract.

   When waves travel from one material or medium to another they change speed – this is refraction. The change of speed causes a change in **direction** unless the wave hits the boundary at a right angle. During refraction the frequency does not change, but the wavelength decreases (if the wave is slowing down) or increases (if the wave is speeding up).

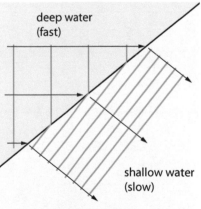

3. All waves **diffract**.

   Waves spread out when they pass through a gap. This is called diffraction.

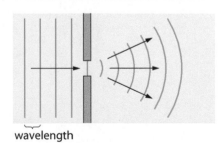

wavelength

Diffraction happens most when the width of the gap is around the same size as the wavelength of the wave. Diffraction also happens when the waves pass an edge (think of it as a narrow gap that only has one side).

Definition

Wavespeed is calculated using:

$$v = f \times \lambda$$

where:

- v is wavespeed in metres per second, m/s
- f is frequency in hertz, Hz
- λ is wavelength in metres, m.

Revision tip

The wave equation needs to be memorised so you can recall it in exams.

Quick test

1. Complete the sentence – waves transfer _____ without transferring _____ .
2. Describe the difference between longitudinal waves and transverse waves.
3. Sketch a wave shape and label the wavelength and the amplitude.
4. What is refraction?
5. What is diffraction?

Supplement

6. When does diffraction happen the most?
7. Write down the wave equation linking wavespeed, frequency and wavelength.

Reflection of light

When light hits a reflecting surface such as a mirror it reflects (bounces off). To measure the angles involved, we draw a construction line at 90° to the surface called a normal line.

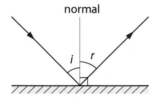

We measure angles from the **incident ray** to the normal line (the angle of incidence) and from the **reflected ray** to the normal line (the angle of reflection). This gives us the law of reflection, which says that:

> the angle of incidence = the angle of reflection

We can use this law to explain how a mirror forms a picture (an 'image') when we look into it.

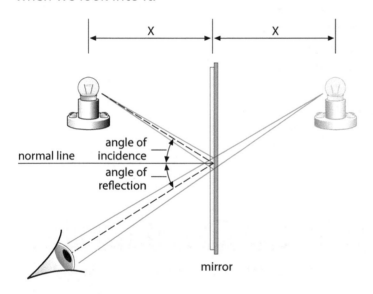

The image in a plane mirror is:

- upright (the right way up)
- laterally inverted (left and right are swapped over)
- as far behind the mirror as the object is in front.

Revision tip

Note that it is a **plane** mirror, not a **plain** mirror! 'Plane' means 'flat'.

Supplement

The image in a plane mirror is virtual. This means that the light only **appears** to come from behind the mirror, it doesn't **actually** come from there – if you were to place a screen behind the mirror there would be no picture on it.

Refraction of light

Refraction is an effect that happens when light crosses the boundary from one medium to another.

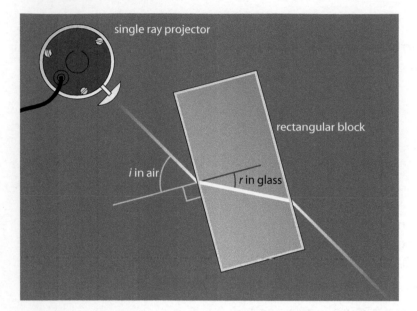

When the light hits the block it slows down. This causes the light to change direction. As it is slowing down, the light is pulled towards the normal line and follows a different path inside the block. As it leaves the block, the light speeds back up again and this speeding up causes the light to move away from the normal line and onto a path parallel to the original path (as it is returning to the air).

As with the plane mirror, the **angle of incidence** is the angle between the ray of light and the normal line as the light hits the boundary. We use the term **angle of refraction** for the angle between the ray of light and the normal line after the light has crossed the boundary.

Supplement

> **Definition**
>
> Refractive index, n, is a number calculated to show how big an effect refraction has at a boundary between two materials. It is calculated in two ways.
>
> - From the change of speed involved:
>   refractive index, n = speed of light in a vacuum (or air) / speed of light in the material
> - From the angles measured at the boundary:
>   refractive index, $n = \sin i \, / \sin r$
>
>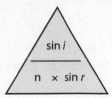

An interesting effect happens if we look more closely at the refraction that happens when the light travels from an optically more dense medium into a less dense medium.

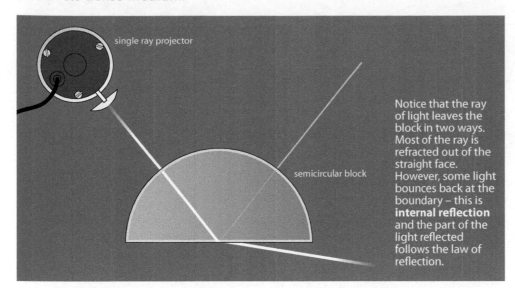

single ray projector

semicircular block

Notice that the ray of light leaves the block in two ways. Most of the ray is refracted out of the straight face. However, some light bounces back at the boundary – this is **internal reflection** and the part of the light reflected follows the law of reflection.

Since the light is speeding up as it exits the block, the light moves away from the normal line. If we gradually increase the angle of incidence, we reach a point where the **emergent ray** exits with an angle of refraction of 90° – it runs along the boundary. When this happens we call the angle of incidence the critical angle.

If we increase the angle of incidence any further, then it is not possible for the ray of light to emerge from the block at all – the light is **totally internally reflected** inside the block.

> ### Revision tip
>
> Remember that for **total internal reflection** to happen, **two** things need to be true – the angle of incidence must be greater than the critical angle **and** the light must be passing from an optically more dense medium to a less dense medium.

Supplement

When light strikes the boundary at the critical angle, c, then the angle of refraction is 90° and $\sin r = 1$. This means that the equation to calculate refractive index becomes:

$$n = 1 / \sin c$$

Total internal reflection allows light to be sent along **optical fibres**.

light beam

Optical fibres can be used to direct light along paths in a range of devices. For example, **endoscopes** use bundles of optical fibres to direct light beams along the digestive tract in humans to allow for examinations and diagnoses without the need for surgery. Optical fibres also allow pulses of light to be sent along them, allowing for coded information to be sent quickly and efficiently – for example in fibre broadband internet communications.

Thin converging lenses

Light refracts at a boundary. By shaping pieces of glass, or other transparent materials, carefully, we can make **lenses** to form images.

The basic action of a converging lens is to cause parallel rays of light to **converge** (come together) to a point, called the **principal focus, F**.

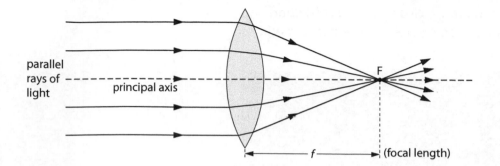

parallel rays of light

principal axis

F

f (focal length)

Revision tip

Converging lenses are also called convex (because of their shape) and positive lenses.

The distance from the lens to the principal focus is called the focal length.

To find the position of the image, draw a **ray diagram** and follow these rules.

1. Draw the object so that its base is on the principal axis (this is usually already done for you).

2. Draw a ray of light from the top of the object through the **centre** of the lens – this ray does not change direction.

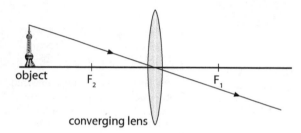

object F_2 F_1

converging lens

3. Draw a ray of light from the top of the object parallel to the principal axis – this ray is refracted through the principal focus.

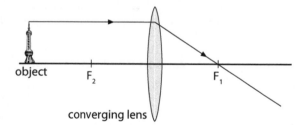

object F_2 F_1

converging lens

4. Draw a ray of light from the top of the object through the principal focus to the lens – this ray is refracted parallel to the principal axis.

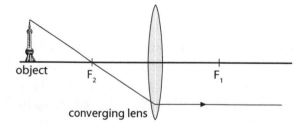

object F_2 F_1

converging lens

5. Look for the point where the three rays meet – this is the position of the image.

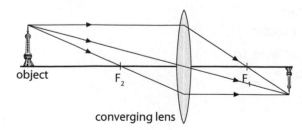

object F_2 F_1

converging lens

Images from lenses can be **enlarged** (larger than the object), the **same size** as the object or **diminished** (smaller than the object). Images can also be **inverted** (upside down) or **upright** (the right way up).

In the example above, the image is inverted and diminished.

Supplement

The size and position of the image changes depending on how far the object is from the lens. In this example the object is closer to the lens than the focal length.

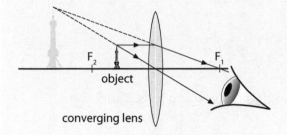

To draw this diagram, follow the rules given above. However, instead of looking for where the rays meet, you need to extend them back to where they **would** meet. This is an example of a virtual image – the rays of light appear to come from there, but don't actually come from that point. The previous example showed a real image – the rays of light actually go to that place; a screen at that position would have a picture on it. When the object is closer than the principal focus the image is **enlarged** and the lens is acting as a **magnifying glass**.

Dispersion of light

When white light is shone into a triangular prism, the different colours change speed by different amounts and so they follow different paths. This results in a **spectrum** of colours emerging from the prism.

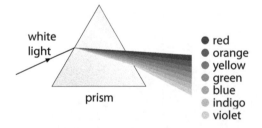

This splitting of white light into the separate colours is called dispersion.

Revision tip

The different colours of light have different **wavelengths**. Light with a single wavelength is called **monochromatic**.

Quick test

1. Draw a diagram to show a ray of light reflecting off a plane mirror. Label the normal, angle of incidence and angle of reflection.
2. State the law of reflection.
3. Describe the properties of the image in a plane mirror.

4. What is a **virtual** image?

5. Draw a diagram to show a ray of light refracting as it passes through a glass block. Label the normal, angle of incidence and angle of refraction.

6. What does refractive index measure?

7. How can refractive index be calculated?

8. What is total internal reflection?

9. When does total internal reflection happen?

10. Draw a diagram to show parallel rays of light passing through a converging lens. Label the principal axis, the principal focus and the focal length.

11. Draw a diagram to show how a lens forms an image of an object when the object is more than twice the focal length from the lens.

12. Draw a diagram to show how a lens forms an image of an object when the object is closer to the lens than the focal length.

13. Dispersion can be shown with a triangular prism. What is dispersion?

Electromagnetic spectrum

The different colours of the spectrum of visible light are part of a larger 'family' of waves called the electromagnetic spectrum.

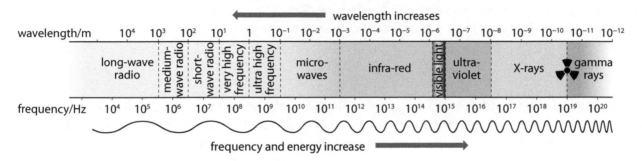

All these waves travel at the same speed in a vacuum (space).

The waves of the electromagnetic spectrum have a variety of **uses** and some **dangers**. The table shows some of them.

> ### Revision tip
>
> Memorise the speed of electromagnetic waves in a vacuum – it is 3.0×10^8 m/s. The speed in air is approximately the same.

Part of spectrum	Uses	Dangers
radio	communication – terrestrial TV signals and radio	
microwaves	mobile phone signals, cooking	microwaves cause water molecules to vibrate with greater energy, so they heat up objects containing water, such as humans
infra-red	remote controls, intruder alarms	infra-red radiation is heat radiation – it can burn
visible light	communication	
ultraviolet	security marking	can damage cells, can cause skin cancers
X-rays	medical imaging, security scanning (for example at airports)	can penetrate the body and damage cells, can cause cancers
gamma	sterilising medical instruments, medical imaging, medical treatment	can penetrate the body and damage cells, can cause cancers

Quick test

1. List the parts of the electromagnetic spectrum in order of increasing frequency.
2. How would the order change if they were listed in order of increasing wavelength?

Supplement

3. Write down the speed of all electromagnetic waves in a vacuum.

4. List one use of each part of the electromagnetic spectrum.
5. Describe how microwaves can be dangerous.
6. Describe how X-rays and gamma radiation can be dangerous.

Sound

Sound is caused by vibrations. Examples include a rule 'twanged' on a bench or a guitar string vibrating.

Sound travels as a longitudinal pressure wave and needs a **medium** (material) for the vibrations to pass along – sound cannot travel through a vacuum.

Supplement

As the pressure wave of the sound passes through the air, the particles will be a little closer together than usual in some places – these are called compressions; and a little further apart in others – these are called rarefactions. Remember back to describing longitudinal waves on a slinky spring as a comparison.

Experiment

To measure the **speed of sound** you can use a sound source, such as a loudspeaker, with two microphones placed directly in front of it but different distances away. Connect the microphones to a fast-response data logger or computer. When a sound is produced, the computer records the time when the sound reached each microphone. If you measure the distance between the microphones, then the speed of sound is calculated from:

speed of sound = distance between microphones / time difference between sound arriving

Alternatively, we can make use of another wave property – reflection. As with all waves, sound waves reflect – producing an **echo**. If we stand a known distance from a suitable surface, such as a large wall, and make a loud sound we can time how long it takes for the echo to arrive back at us. We can then calculate the speed of sound from:

speed of sound = 2 × distance to wall / time between making sound and hearing echo

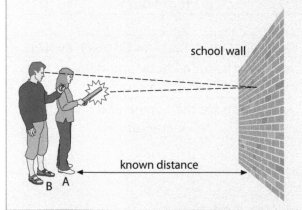

school wall

known distance

B   A

Supplement

Typical values for the speed of sound are:

- steel: 3200 m/s
- water: 1500 m/s
- air: 340 m/s.

Sound travels as a vibration from one particle to the next, so it tends to travel faster in solids because the particles are very close together.

Sound waves have 'everyday' descriptions such as 'loudness' and 'pitch' that we can relate to the words describing waves that were introduced earlier.

A high pitch sound has a high frequency.

A **loud** sound has a large amplitude.

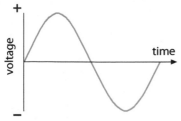

A loud sound of low frequency

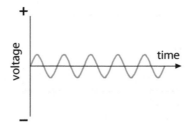

A loud sound of high frequency

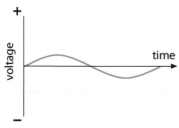

A quiet sound of low frequency

A quiet sound of high frequency

We detect sounds with our ears – the vibrations of the air set the ear drum vibrating and this is transferred into messages sent to the brain. However, human ears cannot respond to all possible sound vibrations.

The **range** of human hearing is from 20 Hz (lowest sounds we can hear) to 20 000 Hz (the highest we can hear). Other animals can hear over different ranges.

Sound with **frequencies above the range of human hearing** is called ultrasound, which has a range of uses such as medical imaging, treating kidney stones and cleaning delicate instruments such as watches.

Quick test

1. Complete the sentence – all sounds are caused by _____ .
2. Explain why sound cannot travel through a vacuum.

Supplement

3. What are compressions and rarefactions?

4. Describe an experiment to measure the speed of sound in air.

Supplement

5. Explain why sound usually travels more quickly in solids compared to gases.

6. How are pitch and loudness related to frequency and amplitude?
7. State the range of human hearing.
8. What is ultrasound?

Exam-style practice questions

1. The diagram below shows water waves on the surface of a ripple tank.

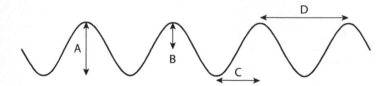

   (a) State which label, A, B, C or D, shows the wavelength of the waves. [1]

   (b) State which label, A, B, C or D, shows the amplitude of the waves. [1]

   (c) The water waves are *transverse* waves. Explain what this means. [2]

   (d) A student uses a ripple tank to demonstrate diffraction.

      (i) Complete the diagram below to show the wave pattern they would observe. [2]

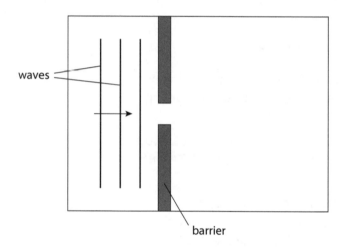

Supplement

      (ii) Describe how the width of the gap affects the amount of diffraction observed. [1]

2. (a) The diagram shows a ray of light incident on a plane mirror.

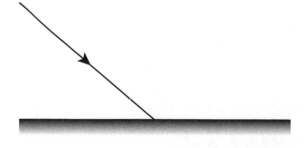

      (i) Complete the diagram by drawing a normal line and the reflected ray. [2]

      (ii) On the diagram, label the angle of incidence and the angle of reflection. [2]

(iii) Complete the following sentence by choosing the correct words.

The angle of incidence is less than / equal to / greater than the angle of reflection. [1]

Supplement

(iv) The image formed by a plane mirror is *virtual*. Explain what this means. [1]

(b) A ray of light is refracted as it enters a glass block from the air.

(i) State what happens during the process of refraction. [2]

Supplement

(ii) The speed of light in air is 3×10^8 m/s and the refractive index of the glass block is 1.5. Calculate the speed of light in the glass block. [2]

(c) An endoscope is a medical device used to observe the digestive tract. It uses total internal reflection to send light along optical fibres.

(i) State the conditions needed for total internal reflection to happen. [2]

(ii) Suggest an advantage of using optical fibres in an endoscope. [1]

3 The electromagnetic spectrum has seven sections.

radio	microwaves	infra-red	visible	ultraviolet	X-rays	gamma

(a) **(i)** Name **two** sections of the spectrum that are used for cooking. [2]

(ii) Name the section of the spectrum that causes a sun tan. [1]

(b) Complete the following sentence by choosing the correct words.

The arrow below the table shows the direction of increasing wavelength / increasing frequency / increasing wavespeed. [1]

Supplement

(c) A radio station broadcasts signals with a wavelength of 1500 m.

(i) State the speed of radio waves in a vacuum. [1]

(ii) Calculate the frequency of this radio station, giving a suitable unit. [3]

4 **(a)** Sound waves are *longitudinal* waves. Explain what this means. [2]

(b) Two students measure the speed of sound in air. They stand 135 m from a wall. One student starts a stopwatch when the second student fires a starting pistol. The student stops the stopwatch 0.8 s later when they hear the echo from the wall. Calculate the speed of sound in air. [3]

Supplement

(c) The speed of sound in steel is 3200 m/s. Use ideas about particles to explain why sound travels more quickly in steel than in the air. [2]

(d) (i) Describe the difference between sound and ultrasound. [2]

(ii) Ultrasound is used in pre-natal scanning during pregnancies. Explain why ultrasound is used rather than X-rays. [2]

Simple phenomena of magnetism

Magnetism is a **non-contact** force that affects magnets and magnetic materials.

A **magnet** is an object that has a magnetic field of its own. A magnetic field is the region of space around a magnet where it can have an effect. A **magnetic material** is a material that is affected by the magnetic field of nearby magnets.

The magnetic field of a **bar magnet** is strongest at the ends of the magnet. These ends are called **poles** and are labelled **north** (N) and **south** (S). The magnetic field is shown by drawing **magnetic field lines**. The diagram shows the magnetic field pattern for a bar magnet.

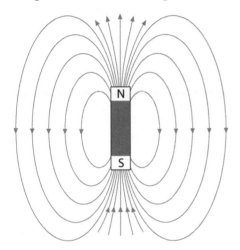

The field lines always point from N to S and never cross (because they show the direction of the force on a N-pole at that point).

Experiment

The shape of a magnetic field can be shown by sprinkling iron filings near the magnet.

To plot the field more accurately, use a **plotting compass**. Marking the position of the compass needle in a variety of places around the magnet will reveal the shape of the field. The direction of the field lines is given by the direction the compass needle points.

Magnetic forces can **attract** (pull together) or **repel** (push apart).

if a ...	is brought up to a ...	then the force will be ...
magnet N-pole	magnet N-pole	repulsion
magnet S-pole	magnet S-pole	repulsion
magnet N-pole	magnet S-pole	attraction
magnetic N or S-pole	magnetic material	attraction

Definition

Two important magnetic materials are **iron** and **steel**. Iron is a **soft** magnetic material – it can be made into a magnet but it does not **stay** magnetised. Steel can also be magnetised, but it **does** keep its magnetism – steel makes **permanent** magnets.

Magnetism is induced in a soft magnetic material when it is brought near a magnet. This means that the magnetic material is temporarily magnetised by the magnetic field of the magnet.

The pole of a permanent magnet always induces the opposite pole in an unmagnetised piece of magnetic material.

Magnetic materials can be magnetised by:

- stroking them in one direction with a magnet
- placing them inside a coil that has a direct current in it
- hammering the material while it is in a magnetic field.

Supplement

Magnets can be **de-magnetised** by:

- heating the magnet
- placing the magnet inside a coil that has an alternating current in it
- hammering the magnet when it is not in a magnetic field.

The magnetic effect of a current

Electromagnets are made by wrapping a wire carrying an electric current around a soft iron core. A magnetic field is produced around the coil (sometimes called a **solenoid**) as shown on the right.

Having a soft iron core means that the magnetism is not permanent – when the current is switched off, the core loses its magnetism. This makes electromagnets very useful in places such as scrapyards for moving old cars around. It is also useful in an electric switch called a relay.

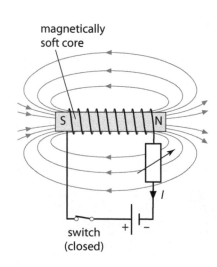

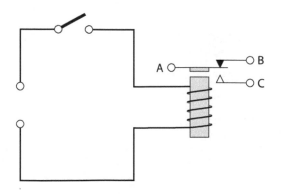

When a current flows in the coil it makes an electromagnet. This attracts the piece of soft iron and connects points A and C. These can be used as a switch to complete a second circuit.

Definition

A single, straight wire carrying an electric current also produces a magnetic field around it. This time the magnetic field forms a circular pattern around the wire.

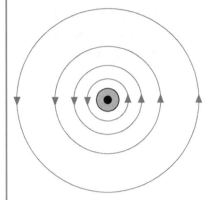

Supplement

In the example, the current is directed towards you out of the page. If the current was directed into the page, the field lines would have the same shape but the arrows would point in the opposite direction.

The direction of the field lines shows the direction of the force on the N-pole of a magnet at that point.

The magnetic field is **stronger** where the field lines are closer together – nearer the wire. The strength of the field can be **increased** by increasing the current.

Quick test

1. What is the difference between a magnet and a magnetic material?
2. To what direction do magnetic field lines point?
3. Explain why magnetic field lines never cross.
4. Describe how to find the shape of a magnetic field around a bar magnet.
5. Iron is a soft magnetic material. What does this mean?
6. State two methods to magnetise a magnetic material.
7. Describe how to make a simple electromagnet.
8. What shape is the magnetic field around a wire that carries an electric current?
9. Describe how a relay works as a switch.

Supplement

10. What information does the closeness of field lines give?

Electrical quantities

Electric charge

Atoms have a central nucleus, which contains **positively charged** protons and neutrons that have no charge, surrounded by electrons that are **negatively charged** (see page 77). **Electric** effects are caused by forces between these charged particles.

Positive charges repel other positive charges and negative charges repel other negative charges. Positive and negative charges attract each other.

Some materials allow electric charges to move through them easily. These materials are called electrical conductors. Metals are all good conductors. Electrical insulators are materials that do **not** allow electric charges to move through them. Non-metals such as wood and plastic are good insulators.

Supplement

Charge is measured in coulombs. The symbol for coulombs is C.

Supplement

In a conductor, some electrons are **delocalised** from the atoms. These electrons can move through the structure.

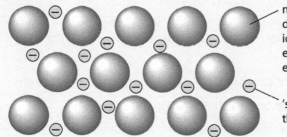

metal atoms (some people describe them as positive ions because they donate electrons into the 'sea' of electrons)

'sea' of electrons (holds the metal atoms together)

In a metal structure the metal ions are surrounded by a cloud or 'sea' of electrons.

Insulators can be given an electrostatic charge by adding or removing electrons. A simple way to do this is by rubbing a plastic rod with a cloth.

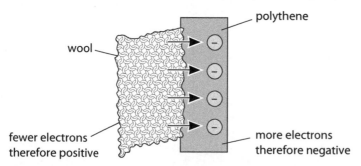

polythene

wool

fewer electrons therefore positive

more electrons therefore negative

Electrostatically charged objects can be detected by their ability to attract small pieces of paper or by deflecting a stream of water running from a tap.

An **electric field** is the region around a charge where it can exert a force. Electric fields are described using **field lines**. These show the direction of the force on a positive charge at that point.

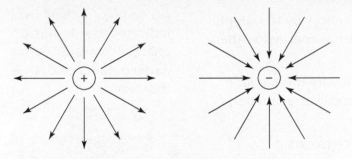

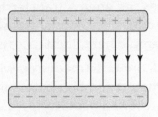

A Electric field round a point charge

B Electric field around two parallel oppositely charged plates

Objects can also be charged by **induction**. For example, if a balloon is rubbed on clothing it can 'stick' to a wall or ceiling. If the balloon has a negative charge (extra electrons) it will repel electrons from the surface of a ceiling if it is placed close to it. This leaves the surface of the ceiling slightly positive – it has an induced charge.

> **Revision tip**
>
> The electric field around a charged, conducting sphere is the same shape as the field around a point charge.

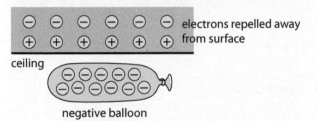

ceiling

electrons repelled away from surface

negative balloon

Current

Moving charge from one place to another is called an electric current and the pathway is called an **electric circuit**. In wires, it is **electrons** that flow to transfer the charge.

Electric current is measured with an **ammeter** and is measured in **amperes**, **A** (often shortened to 'amps'). Ammeters can be **digital** (giving a display directly as a number) or **analogue** (where a needle moves across a scale).

Supplement

Definition
Electric current is calculated using:

$$I = Q / t$$

where:

- I is the current in amps, A
- Q is the charge transferred in coulombs, C
- t is the time taken in seconds, s.

By **convention** (an agreed system), an arrow is drawn on circuit diagrams showing current flowing from the positive terminal to the negative terminal. In circuits using wires, the electrons actually travel in the opposite direction, from the negative terminal to the positive.

Electromotive force, potential difference and electrical working

Energy is transferred when charge flows around an electric circuit. Electromotive force (e.m.f.) and potential difference (p.d.) measure how much energy is transferred for each unit of charge that flows.

E.m.f. describes the energy transferred **into** electrical energy by the **supply** while p.d. describes the energy transferred from electrical energy to **other forms** (for example heat, light) across a component.

E.m.f. and p.d. are both measured with a **voltmeter** and are measured in **volts, V**. Voltmeters can be digital (giving a display directly as a number) or analogue (where a needle moves across a scale).

The process of transferring energy from a power source to circuit components and then into the surroundings is called electrical working.

Supplement

An e.m.f. or p.d. of 1 volt indicates that 1 joule of energy is transferred for each coulomb of charge that flows.

$$1V = 1J/C$$

Revision tip

Remember that 'doing work' means the same as 'transferring energy'.

Supplement

Calculating electrical energy

The energy transferred in a circuit is calculated using:

$$E = I \times V \times t$$

where:

- E is the energy transferred in joules, J
- I is the current in amps, A
- V is the p.d. in volts, V
- t is the time taken in seconds, s

and the energy transferred per second (the power) is calculated using:

$$P = I \times V$$

where:

- P is the power in watts, W
- I is the current in amps, A
- V is the p.d. in volts, V.

Resistance

Resistance measures how difficult it is for an electric current to flow through a component. The higher the resistance, if the potential difference is constant, the lower the current.

Definition

Resistance is calculated using:

$$R = V / I$$

where:

- R is the resistance in ohms, Ω
- V is the p.d. in volts, V
- I is the current in amps, A.

Experiment

This circuit can be used to measure the resistance of a component.

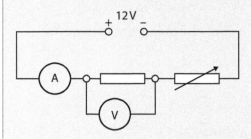

A simple method is to measure the current and potential difference and then calculate the resistance from:

$R = V / I$

An improved method is:

* measure the current and potential difference for several settings of the variable resistor
* plot a graph of potential difference against current
* the resistance is equal to the gradient of the graph.

Supplement

The way that current and potential difference are linked in a component is called the **current–voltage characteristic** for the component. These are often shown as graphs – two important ones are shown below.

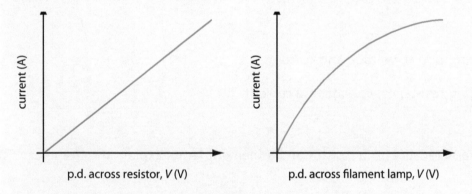

On the left is the graph for an **ohmic** resistor – this is a resistor where the current is proportional to the potential difference and the resistance stays **constant**. The graph on the right shows the characteristic curve for a **filament lamp** – the graph curves, showing that the resistance **increases** at higher currents. This is because the **temperature** rises, making it more difficult for electrons to pass through (as the lattice ions vibrate more vigorously).

Revision tip

'Ohmic resistors' get their name because they follow **Ohm's law**, that the current is proportional to the potential difference.

The resistance of a wire depends on its physical dimensions:

* resistance **increases** as **length** increases
* resistance **increases** as diameter **decreases**

The mathematical relations linking resistance to physical dimensions are:

* resistance is **proportional** to length
* resistance is **inversely proportional** to **cross-sectional area**

Note that if the diameter of a wire **doubles**, the cross-sectional area is **four times larger**, as the area depends on the radius squared.

Quick test

1. Describe the difference between electrical conductors and electrical insulators.
2. Which kind of materials can be given an electrostatic charge?
3. Describe how to detect electrostatically charged objects.

Supplement

4. How can an object be charged 'by induction'?

5. What is electric current measured with?
6. What is the difference between the display of a digital meter and an analogue meter?

Supplement

7. Write down the equation linking current, charge and time.

8. Describe one similarity and one difference between e.m.f. and potential difference.
9. What is 'electrical working'?

Supplement

10. Write down the equation linking power, voltage and current.

11. Write down the equation linking resistance, voltage and current.

Supplement

12. Sketch the current–voltage characteristics for a resistor and a filament lamp. Explain the shape of the two graphs.

13. How does the length and diameter of a wire affect its resistance?

Circuit diagrams

Electric circuits are drawn using a set of standard symbols.

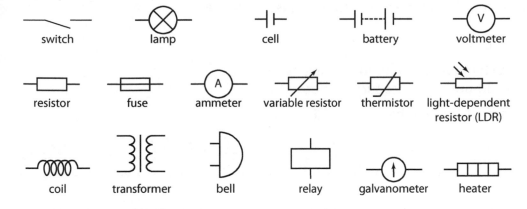

switch	lamp	cell	battery	voltmeter
resistor	fuse	ammeter	variable resistor	thermistor	light-dependent resistor (LDR)
coil	transformer	bell	relay	galvanometer	heater

Supplement

The circuit symbol for the diode is also required.

diode

Series and parallel circuits

In a series circuit, the components are connected one after another, like programmes in a TV series. In a parallel circuit, the components are connected side by side, like parallel lines.

	Series	Parallel
Circuit diagram		
Appearance of the lamps	Both lamps have the same brightness, both are dim.	Both lamps have the same brightness, both are bright.
Current	The current is the same at all points around the circuit.	The current from the source is larger than the current in each branch. **Supplement** The current from the source is the sum of the current in each branch.
Advantages and disadvantages in a lighting circuit	The lamps cannot be switched off independently. If one lamp 'blows', the circuit is incomplete so both go out. The battery will last longer.	The lamps can be switched off separately by placing switches in the separate branches. If one lamp 'blows', the other remains on. The battery will drain more quickly.
Combined resistance	The combined resistance of the circuit is found by **adding together** the resistances of each component.	The combined resistance of the circuit is **less than** the resistance of any component on its own.

Supplement

In a series circuit, adding together the potential differences around the circuit will give a voltage **equal to** the e.m.f. of the source. This is due to the **conservation of energy** – check the definitions of e.m.f. and p.d. to remember the connection between voltage and energy.

Similarly, connecting sources in series, for example connecting batteries together, gives a total e.m.f. that is simply the sum of the separate e.m.f.s.

The equation to calculate the combined resistance of resistors in parallel is:

$$1 / R = 1 / R_1 + 1 / R_2$$

If there are just two resistors, this becomes

Combined resistance = (product of the resistances) / (sum of the resistances)

Action and use of circuit components

Components with a resistance that **changes** can be useful in circuits that sense the surroundings or control additional devices such as motors.

Some important components are shown in the table.

Component	What it does
potentiometer	A variable resistor that can be used to split a voltage into two parts, making a potential divider.
thermistor	A component where the **resistance decreases** as its **temperature increases**.
light-dependent resistor (LDR)	A component where the **resistance decreases** as the **light intensity increases**.
relay	A switch where a coil is magnetised by an electric current, attracting the switch contacts together.
Supplement	
diode	A diode allows electrons to flow **in one direction only**. This is used in the **rectifier** – see below.

Supplement

Worked example

Question

In the circuit below, the relay contacts are connected to an external light bulb. Describe the operation of the circuit as the light level decreases.

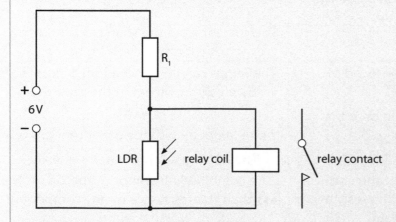

Answer

The LDR and the resistor R_1 form a potential divider 'sharing' the potential difference provided by the source. As the light level decreases, the resistance of the LDR **increases**, so it takes a bigger share of the p.d. When the share of the p.d. is large enough, the relay coil is magnetised enough to attract the contact together, switching on the external light bulb.

In a **rectifier**, a **diode** is used to convert alternating current (a.c.) to direct current (d.c.).

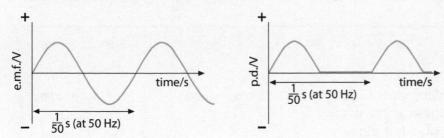

In a.c. the electrons reverse direction in the wire ('alternate'). Because the diode will not allow electrons to pass in the 'reverse' direction, only half the cycle passes through. This is called half-wave rectification.

Quick test

1. Draw the circuit symbols for an ammeter, a resistor, a thermistor and a light-dependent resistor.
2. Explain why the current is the same at all points in a series circuit.
3. Explain why the current can be different in different parts of a parallel circuit.
4. Describe how to calculate the effective resistance of three resistors in series.
5. Describe how the resistance of a thermistor is affect by temperature.

Supplement

6. Describe the action of a diode.

7. What is a 'potential divider'?

Supplement

8. What is a 'rectifier'?

Digital electronics

Supplement

Electrical signals can be sent using analogue or digital coding. In analogue signals, the voltages can have any value (within the limits set by the supply), but digital signals are simply 'high' or 'low' (often described using the numbers '1' for high and '0' for low).

Logic gates are simple circuits that operate with digital signals and are designed to provide an **output signal** that depends on the combination of the **input signals**. The operation of a logic gate is described in a truth table.

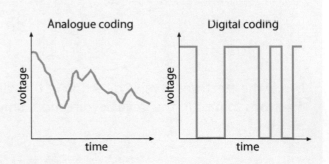

Name of logic gate	Circuit symbol	Truth table

NOT

input ──▷o── output

Input	Output
0	1
1	0

AND

input A ──┐
input B ──┘ ──D── output

Inputs		Output
A	B	
0	0	0
0	1	0
1	0	0
1	1	1

OR

input A ──┐
input B ──┘ ──D── output

Inputs		Output
A	B	
0	0	0
0	1	1
1	0	1
1	1	1

NAND

input A ──┐
input B ──┘ ──Do── output

Inputs		Output
A	B	
0	0	1
0	1	1
1	0	1
1	1	0

NOR

input A ──┐
input B ──┘ ──Do── output

Inputs		Output
A	B	
0	0	1
0	1	0
1	0	0
1	1	0

Worked example

A house owner wants to use a motor that will bring in a washing line when it is raining or when it is cloudy and not windy.

The following circuit will do this.

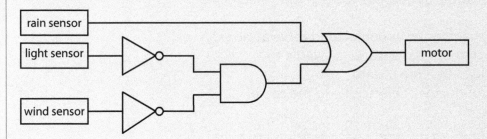

A rain sensor, a light sensor and a wind sensor are used. Notice that NOT gates are used with two sensors (cloudy = NOT light). These two are **both** required, so an AND gate is used for this condition. The OR gate allows the motor to be ON when **either** of the conditions stated in the instruction are true

Quick test

Supplement

1. Draw graphs to show the difference between digital signals and analogue signals.
2. What is a 'logic gate'?
3. What is a 'truth table'?
4. Draw the truth tables for an AND gate and an OR gate.

Dangers of electricity

Electricity can be hazardous. Faults can cause electric shock or electrocution as well as starting fires or damaging appliances.

Typical hazards include:

- damaged insulation or cables – these can expose wires, risking electric shock
- overheating cables – either by having the wrongly rated cable attached or by coiling up a cable (reducing the surface area available to dissipate the heat) – this can lead to a fire hazard
- damp conditions – water is a conductor of electricity, so the risk of connection to a user is increased.

To reduce the risks associated with these hazards, safety features are built in to mains appliances.

Fuses protect a circuit from the effects of the current being too high. Individual appliances contain a fuse and there are also fuses protecting the supply circuits themselves. If the current gets above a stated value the fuse is designed to be the part of the circuit that overheats and **melts** before any other part. In this way, the circuit is broken and the current is switched off.

Fuses are typically supplied with standard 'ratings' – the maximum value of current above which the fuse will melt. The fuse in a circuit **must** have the correct rating – it must be the **lowest value above** the required current. For example, fuses are usually supplied with ratings of 1 A, 3 A, 5 A and 13 A. An appliance designed to run at 3.4 A should have a 5 A fuse fitted.

Circuit breakers carry out the same job as a fuse. This time, rather than melting to break the circuit, the circuit breaker switches ('trips') a spring-loaded switch to stop the circuit working. Two advantages of using a circuit breaker rather than a fuse are:

- they can act more quickly
- they can be 'reset' easily once the fault in the circuit has been corrected (a fuse has to be replaced).

Earth wires are designed to protect the **user** from electric shock. They are only required when the appliance has a **metal case** (which can conduct through to the user).

If the appliance is working correctly, energy is delivered through the live (L) and neutral (N) wires and the earth wire (E) is not used.

If a fault occurs and the metal case becomes 'live':

- the earth wire provides a **low-resistance path** to the earth
- so a **large current** flows along that path
- which overheats and melts the **fuse**
- switching the appliance off and making it safe to touch.

> ## Revision tip
>
> The fuse does NOT 'stop the current getting too high'; it simply breaks the circuit and turns it off if the current does exceed the rated value.

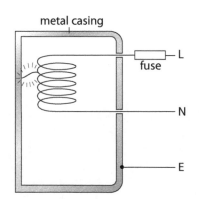

metal casing

L
fuse
N
E

Quick test

1. State three electrical hazards.
2. Describe how a fuse protects an appliance.
3. An appliance is designed to operate at 2.3 A. Explain why a 3 A fuse would be better than a 13 A fuse in this appliance.
4. What advantages does a circuit breaker have over a fuse?
5. Why do appliances with plastic cases not need an earth wire?
6. Describe how an earth wire protects the use of an appliance.

Electromagnetic effects

Electromagnetic induction

When there is relative motion between a conductor (such as a wire) and a magnetic field, an e.m.f. (voltage) is induced (created) across the ends of the conductor.

To show this effect, a wire can be moved in the magnetic field of a permanent magnet.

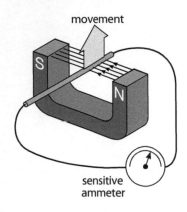

movement

sensitive ammeter

> **Revision tip**
>
> 'Induced' in this context means 'without contact'. It has the same meaning when looking at **induced magnetism** or charging by induction.

To **increase** the magnitude of the e.m.f. induced:

- increase the strength of the magnetic field
- increase the speed of the motion (of the conductor or the magnetic field)
- wrap the conductor into a coil shape – this increases the amount of conductor cutting through the field at any one time.

> **Revision tip**
>
> An e.m.f. (voltage) is **always** induced in this situation. A **current** can flow only if there is a **complete circuit**.

Supplement

When an e.m.f. is induced, the **direction** of the e.m.f. (which end of the conductor is positive and which end is negative) can be found using **Fleming's right-hand rule**.

If the thumb and first two fingers of the **right** hand are set at right angles to each other, then lining up the **thumb** with the direction of the **motion** and the **first finger** with the **magnetic field** (N to S) leaves the **second finger** pointing in the direction of the **e.m.f.** (from positive to negative).

> **Revision tip**
>
> Do not confuse this with **Fleming's left-hand rule**, which describes the direction of the force caused during the **motor effect** (see page 72).

a.c. generator

A battery produces a direct current (d.c.) – the electrons move as a steady stream, always moving from the negative terminal of the source round to the positive terminal.

An alternating current (a.c.) is different. The electrons move backwards and forwards – they **alternate** their direction. For mains electricity, the motion is repeated regularly at **mains frequency**, which is 50 Hz in many countries, although it does vary.

a.c. is used for mains electricity as there is **less energy wastage** when transferring energy over long distances (see page 72).

The diagram shows a simple a.c. generator and the e.m.f. (voltage) produced by it over a period of time.

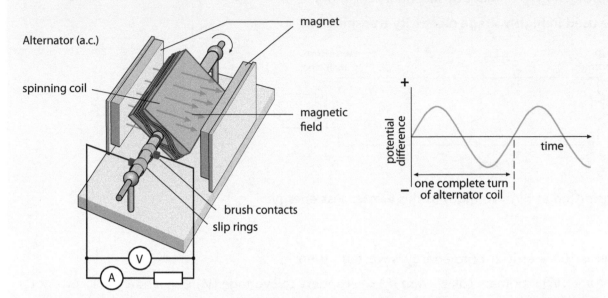

When the coil spins, the wire cuts through the magnetic field and an e.m.f. is induced. The wire is a **coil** to increase the magnitude of the e.m.f. The **slip rings** are contacts at the end of the coil that rest against the **brushes** which in turn connect to an external circuit.

On the graph, the peaks (top and bottom) are formed when the coil is horizontal between the magnetic pole – this is where the coil is cutting through the field lines the fastest. The induced voltage is zero where the coil is vertical.

> **Revision tip**
>
> Do not confuse the action of a **generator** with that of a **motor** (see page 73). The construction is very similar, but in the generator an external energy supply is needed to make the coil spin.

Transformer

Transformers are electrical devices that are designed to **increase** or **decrease voltages**.

Transformers consist of two coils of insulated wire wrapped around a soft iron core.

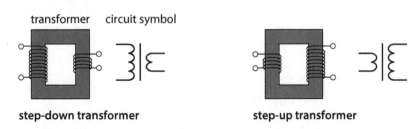

step-down transformer   step-up transformer

When an **alternating** voltage is applied to the **primary** coil it creates a **changing** magnetic field around it. This magnetic field **cuts through** the **secondary** coil, **inducing** a voltage in it. The soft iron core magnetises to increase the strength of the field.

In a **step-up** transformer, the secondary voltage is bigger than the primary voltage; in a **step-down** transformer, the secondary voltage is the smaller.

The relationship between turns and voltages is:

$$\frac{\text{primary coil voltage } (V_p)}{\text{secondary coil voltage } (V_s)} = \frac{\text{number of primary turns } (n_p)}{\text{number of secondary turns } (n_s)}$$

Transformers are used in **high-voltage electricity transmission**.

> **Revision tip**

Remember to refer to 'number of turns' on a coil, not 'number of coils'.

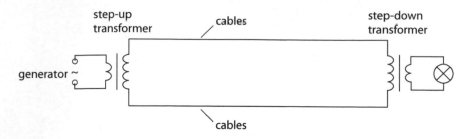

Electricity is transmitted at high voltages as this **wastes less energy**.

Supplement

If a transformer is 100% efficient (no energy is wasted), then:

Primary coil voltage (V_p) × primary coil current (I_p) = secondary coil voltage (V_s) × secondary coil current (I_s)

This means that sending electricity at high voltages means sending it at **low current**. Lower current means there will be less resistance heating in the cables, meaning that less energy is wasted as heat.

Force on a current-carrying conductor

When a wire carrying an electric current is placed in a magnetic field and at right angles to it, the wire experiences a **force** at right angles to both the field and the direction of the current.

The **direction** of the force is **reversed** if:

- the direction of the **current** is reversed
- the direction of the **magnetic field** is reversed.

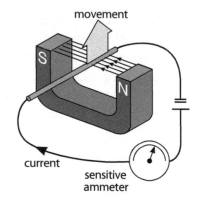

Supplement

When a force acts on a current-carrying conductor, the **direction** of the force can be found using **Fleming's left-hand rule**.

If the thumb and first two fingers of the **left** hand are set at right angles to each other, then lining up the **first finger** with the **magnetic field** (N to S) and the **second finger** with the direction of the **current** (from positive to negative) leaves the **thumb** pointing in direction of the **force**.

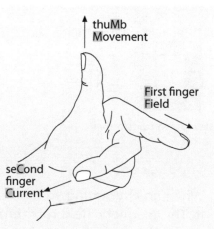

> **Revision tip**

Do not confuse this with **Fleming's right-hand rule**, which describes the direction of the e.m.f. caused during the **electromagnetic induction** (see page 70).

A **beam of charged particles**, such as an electron beam or a proton beam, will also experience a force in a magnetic field – think of the particle beam as an electric current without the wire.

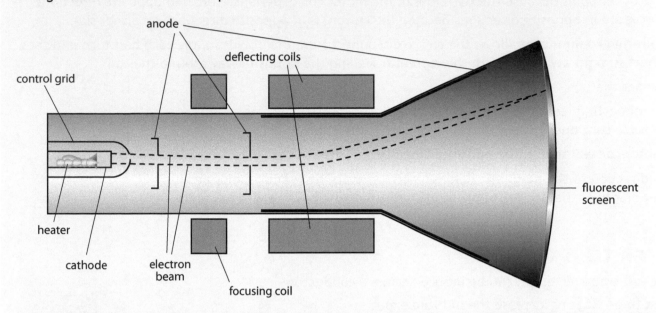

In this example, an electron beam is produced by the cathode and then deflected by the magnetic field produced by the coils.

d.c. motor

The diagram shows a simple direct current (d.c.) motor.

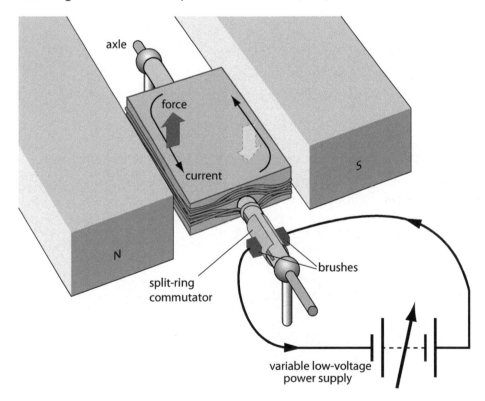

The coil experiences a **turning effect** – the motor coil **spins**.

The effect is **increased** by:

- increasing the **number of turns** on the coil
- increasing the **current**
- increasing the **strength** of the **magnetic field**.

> **Revision tip**

Do NOT say 'use a bigger magnet' when you mean 'increase the magnetic field strength'!

The motor coil spins because the two sides of the motor coil experience forces **in opposite directions**. The forces are in opposite directions because the **current** is in opposite directions on either side.

The **split-ring commutator** allows the coil to disconnect from the power supply each half turn and then reconnect straight away, but with the current travelling the **opposite** way around the coil.

This means that:

- the connecting leads are not tied in a knot as the motor coil spins
- the force continues to spin the coil in the same direction.

The motor coil will spin in the **opposite direction** if:

- the direction of the magnetic field is reversed (swap the N and S-poles around)
- the polarity of the power supply is reversed (swap the positive and negative around).

Quick test

1. Describe how an e.m.f. can be induced across a conductor.
2. List three ways to increase the induced e.m.f.
3. What is the difference between direct current (d.c.) and alternating current (a.c.)?

4. Sketch a graph to show the e.m.f. produced by an a.c. generator as it spins.
5. What is the purpose of the slip rings on an a.c. generator?

6. What is a transformer designed to do?
7. Describe the difference between a step-up transformer and a step-down transformer.
8. Why are transformers used in high-voltage electricity transmission?
9. Describe how a magnetic field affects a current-carrying conductor.

10. State two ways to change the direction of the force on a current-carrying conductor in a magnetic field.
11. Explain how Fleming's left-hand rule gives the direction of the force.

12. Explain why the coil in a simple d.c. motor experiences a turning effect.

13. State two ways to make a simple d.c. motor spin in the opposite direction.

1. A student experiments with a bar magnet.

(a) On the diagram below, draw the magnetic field pattern around a bar magnet. [3]

| N | S |

(b) Describe how the student could use a plotting compass to show the shape of the magnetic field around a bar magnet. [3]

(c) The student uses the magnet to attract a piece of metal. Explain why the student **cannot** be sure if the second piece of metal is a magnet. [2]

(d) The student makes an electromagnet by passing an electric current through a coil wrapped around an iron nail.

   (i) State **two** factors that affect the strength of an electromagnet. [2]

   (ii) State **one** advantage of using an iron core in an electromagnet. [1]

2. A student investigates how the length and diameter of a wire affects its resistance.

(a) Describe how the student should carry out the investigation, making use of an ammeter and a voltmeter.

   You should include:

   • a circuit diagram

   • the measurements the student should make

   • how the student would use their measurements to calculate the resistance [6]

Supplement

(b) The student expects the resistance of the wire to be inversely proportional to its cross-sectional area. What does the term *inversely proportional* mean? [2]

(c) The student should disconnect the circuit between readings. Suggest why this should be done. [2]

3 An electric-bar fire has two heating elements.

(a) The heating elements are connected to the electrical supply in parallel. Give **two** advantages of connecting the heating elements in parallel. [2]

Supplement

(b) The power rating of the heater is 1200 W and it is connected to a 230 V supply. Show that the current in the heater is just over 5 A when it is operating. [2]

(c) (i) State which of these fuses – 1 A, 5 A, 7 A or 13 A should be fitted to the heater. [1]

(ii) Give a reason for your answer. [1]

(d) The heater is fitted with an earth wire. Describe how an earth wire acts as a safety feature in the heater. [4]

4 Transformers are used in the distribution of mains electricity.

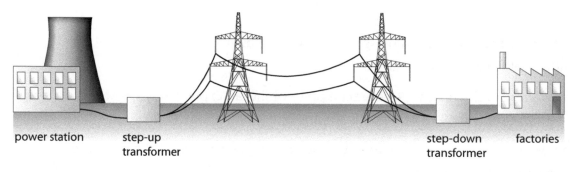

power station   step-up transformer   step-down transformer   factories

(a) What is the purpose of a step-up transformer? [1]

Supplement

(b) Explain why electricity is transmitted at high voltages. [3]

(c) The diagram shows a simple transformer.

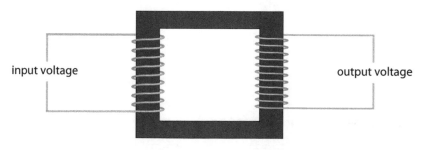

input voltage   output voltage

(i) On the diagram, label the iron core and the primary coil. [2]

(ii) The primary coil has 100 turns and the secondary coil has 400 turns. Calculate the output voltage if the input voltage is 12 V. [3]

Atoms are very small particles.

Atoms have a central nucleus, which is **positively charged**, surrounded by electrons, which are **negatively charged**.

Supplement

Evidence that the atom has a nucleus (which is why we call this the 'nuclear atom') comes from an experiment carried out in the early 1900s.

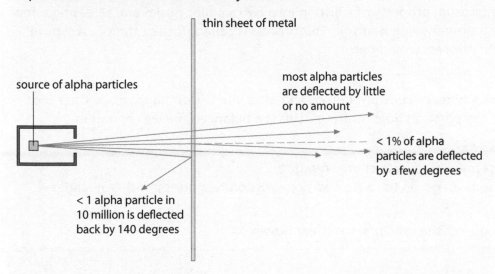

In this experiment, α-particles were fired at a thin, gold target. Almost all the α-particles passed straight through, as expected. However, a small number were deflected through large angles. Because the overall size and mass of the atoms was already known, the only possible explanation for this result was that there must be a very **dense** region within the atom – almost all of the mass of the atom must be concentrated in a small central region, which we now call the nucleus. As it was already known that electrons were negatively charged and that atoms overall were usually neutral, this central nucleus had to have a **positive** charge.

The nucleus itself is made of smaller particles called **protons** and **neutrons**.

The table shows the key properties of the three particles that make up atoms.

Name	Relative charge	Relative mass	Location
proton	+1	1	nucleus
neutron	0	1	nucleus
electron	-1	0.0005	orbits nucleus

The atoms of the different elements have different numbers of protons and neutrons in their nuclei. A particular type of nucleus is called a nuclide. We use **nuclide notation** to describe a nucleus:

$$^{A}_{Z}X$$

- X is the symbol for the chemical element, for example Li for lithium.
- Z is the proton number, which gives the number of protons in the nucleus; in a neutral atom, this is the same as the number of electrons orbiting the nucleus.

- *A* is the nucleon number, which is the total number of protons **and** neutrons in the nucleus.

Atoms of the same element **always** have the same number of protons – for example, uranium **always** has 92 protons. However, the number of neutrons is not always the same. For example, uranium atoms usually have 146 neutrons in the nucleus, but can have any number of neutrons from 135 up to 148. Atoms with the same number of protons but different numbers of neutrons are called isotopes.

Supplement

Nuclear fission

A few heavy nuclides have the unusual property of splitting into two smaller nuclei and releasing a few neutrons when they are hit by a slow-moving neutron. This process is called nuclear fission. A typical **nuclear equation** to describe the process would be:

$$^{235}_{92}U + {}^{1}_{0}n = {}^{137}_{56}Ba + {}^{97}_{36}Kr + {}^{1}_{0}n + {}^{1}_{0}n + Q$$

Note that the number of protons remains constant before and after the fission happens, as does the number of neutrons – none of the particles goes 'missing'. This is a **balanced** nuclear equation.

Nuclear fission is important because:

- a lot of energy is released (represented as Q in the equation)
- the neutrons that are released can go on to cause additional fission reactions in other nuclides – causing a **chain reaction**.

Nuclear fission chain reactions power the reactors in nuclear power stations and nuclear submarines.

Nuclear fusion

Nuclear fusion is a process where **small** nuclei, such as hydrogen, join together ('fuse') to make heavier nuclei, such as helium. In this process, even greater amounts of energy are released. This is the process that powers stars.

> **Revision tip**
>
> Be careful with your spelling of 'fission' and 'fusion' – if you write 'fision' or 'fussion' it will not be clear what you are referring to.

Quick test

1. The nucleus of an atom has what type of charge?
2. Electrons have what type of charge?

Supplement

3. How did the large-angle deflection of alpha particles provide evidence for a small nucleus in an atom?

4. State the relative charges and masses of protons, neutrons and electrons.
5. When writing symbols for an atom, what do the letters X, A and Z represent?
6. What are isotopes?

Supplement

7. What is nuclear fission?
8. What is a chain reaction?
9. What is nuclear fusion?

Detection of radioactivity

The nuclei of some atoms are **unstable**. This means that the nucleus cannot remain as it is forever – it will emit (give out) particles or waves so that it becomes more stable and able to remain with a fixed combination of protons and neutrons from then on. Atoms that have an unstable nucleus are called radioactive and there are three types of particles and waves ('ionising radiation') that can be emitted. Radioactive emission is a random process, which means that we cannot say exactly **which** atom will emit radiation ('**decay**') next or **when** a particular atom will decay.

There is always some ionising radiation around us, caused by radioactive atoms in the environment (such as in rocks or the air) and by artificial sources (such as nuclear reactors or some medical equipment). The ionising radiation that is always around us is called background radiation.

When studying radioactive materials, readings should always be taken for background radiation and these values should be subtracted from the overall measurements – this makes sure you are only studying the radioactive source.

Characteristics of the three kinds of emission

The three possible types of ionising radiation are called **α-particles**, **β-particles** and **γ-rays**. They are called 'ionising' because they can *ionise* (knock electrons out of) atoms that they collide with. They can be detected with photographic film or with a Geiger–Muller tube connected to a counter or ratemeter.

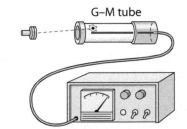

G–M tube

The three types of ionising radiation have different properties.

	alpha (α)	beta (β)	gamma (γ)
Description	A positively charged particle, identical to a helium nucleus (two protons and two neutrons)	A negatively charged particle, identical to an electron	Short-wavelength electromagnetic radiation; electrically neutral
Penetration	4–10 cm of air; stopped by a sheet of paper	About 1 m of air; stopped by a few mm of aluminium	Almost no limit in air; intensity greatly reduced by several cm of lead or several metres of concrete
Ionising power	High	Medium	Low
Supplement			
Effect of electric fields	Positively charged, higher mass particles, so deflected towards negative charges	Negatively charged, low mass particle so deflected considerably towards positive charges	Uncharged, so not affected at all
Effect of magnetic fields	Positively charged, so deflected by magnetic fields – deflection is in the opposite direction to beta radiation	Negatively charged, so deflected by magnetic fields – deflection is greater than for alpha radiation and in the opposite direction	Uncharged, so not affected at all

Alpha particles are the larger of the particle emissions so they collide with atoms in their path easily, giving them the highest ionising ability. This means that they will do the most 'damage' to living cells, for example. However, because they collide so easily they have a low penetrating ability. Beta particles are much smaller so they do penetrate further, but are less likely to be able to cause ionisation. Gamma rays are the most penetrating, so they lose little energy through collisions with atoms and have the lowest ionising ability.

Radioactive decay

When a nucleus emits α-, β- or γ-radiation, the nucleus becomes more stable – this process is called **radioactive decay**.

For α or β decay, the number of protons in the nucleus changes, so the atom becomes an atom of a different element – remember that atoms of a particular element always have the same number of protons. For example, when a nucleus of $^{14}_{6}$C decays it becomes a nucleus of $^{14}_{7}$N.

Supplement

The table shows how the proton number and nucleon number change in each type of radioactive decay.

Nucleus emits	Nucleon number	Proton number
alpha particle	decreases by 4	decreases by 2
beta particle	does not change	increases by 1
gamma ray	does not change	does not change

We can use these ideas to describe radioactive decay with **nuclear equations**, in the same way that we described nuclear fission.

Alpha decay – the nucleus emits an α-particle (2 protons and 2 neutrons)

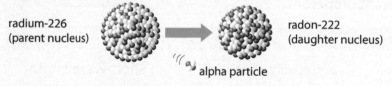

radium-226
(parent nucleus)

radon-222
(daughter nucleus)

alpha particle

$$^{226}_{88}\text{Ra} \rightarrow {}^{222}_{86}\text{Rn} + {}^{4}_{2}\alpha$$

Beta decay – a neutron changes into a proton in the nucleus

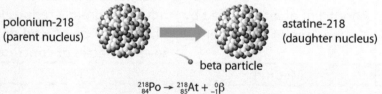

polonium-218
(parent nucleus)

astatine-218
(daughter nucleus)

beta particle

$$^{218}_{84}\text{Po} \rightarrow {}^{218}_{85}\text{At} + {}^{0}_{-1}\beta$$

Gamma decay – the nucleus emits a γ-ray to get rid of excess energy

gamma ray

Half-life

Radioactive materials emit ionising radiation to make the nuclei more stable. Consequently, over time radioactive materials become less radioactive. Since radioactive decay is a **random process**, we cannot say exactly how radioactive a sample will be at any time, but we can say, on average, how long a sample will take to lose **half** its activity – this is called the half-life of the isotope.

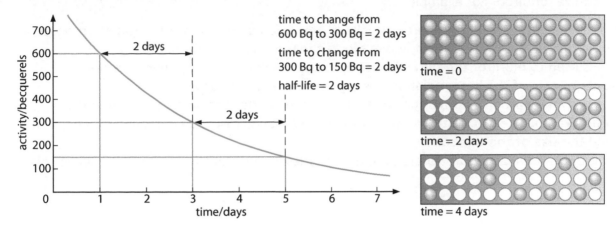

time to change from 600 Bq to 300 Bq = 2 days

time to change from 300 Bq to 150 Bq = 2 days

half-life = 2 days

The half-life is 2 days. Half the number of radioactive nuclei decays in 2 days.

Radioactive materials have a wide range of half-lives, from tiny fractions of a second to many millions of years. This has important consequences for using and storing radioactive materials.

To calculate the half-life from a graph, pick a value from the *y*-axis and draw across to the line and then down to the *x*-axis to find the matching time. Repeat this process, but starting on the *y*-axis at a value **half** of the one you previously chose. The difference between the two times on the *x*-axis is the half-life.

> ### Revision tip
>
> Be careful – the data you are working with may still include the background radiation reading – this should be subtracted so you are only working with data from the radioactive source.

Safety precautions

Radioactive materials emit ionising radiation. The radiation can damage living cells, causing cells to die or divide abnormally. This means they need to be used and stored carefully to avoid unnecessary exposure to the body.

Safety precautions include:

- use tongs to handle radioactive sources, never pick them up directly
- never point radioactive sources at living tissue
- keep radioactive sources in lead-lined containers and lock them away when not in use
- measure background radiation regularly to ensure there is no contamination
- minimise the time when sources are in use
- wear rubber gloves to ensure any spillage does not get on the skin.

For people who work regularly with radioactive sources, for example in hospitals, additional measures could be taken, such as:

- wear lead-lined clothing for additional protection
- monitor exposure to radiation by wearing a **film badge**.

Uses of radioactive isotopes

Radioactive isotopes have a variety of uses. The isotope chosen will depend on a combination of the properties and half-life required. For example:

- Smoke alarms often contain an alpha source, usually $^{241}_{95}$Am. As an alpha source, the radiation poses no threat to the people below, but when smoke blocks the radiation the alarm is triggered.
- Beta radiation is used to monitor the thickness of paper being rolled – if the paper thickness changes, the count rate of the beta radiation passing through changes as well.
- Gamma radiation can be used to destroy bacteria – this is used to sterilise medical instruments.
- Ionising radiation can be used to destroy tumour cells – the particular isotope chosen would depend on the site and the size of the tumour.
- Radioactive materials can be used as tracers to find blockages in underground pipes or blood vessels in the body.

Quick test

1. Some atoms are *unstable*. What does this mean?
2. Radioactive emission is a *random* process. What does this mean?
3. The emissions from radioactive materials are called *ionising*. What does this mean?
4. What is background radiation? List two sources of background radiation.
5. Name a detector of ionising radiation.
6. Describe the properties of alpha, beta and gamma radiation in terms of **(a)** penetrating power, **(b)** range in air, **(c)** ionising power.

Supplement

7. Explain why alpha particles are highly ionising.

8. Explain why alpha or beta decay causes an atom to become a different element.
9. Each radioactive isotope has a *half-life*. What does this mean?
10. How could you find the half-life of an isotope from a graph showing the activity of a sample over a period of time?
11. Why is ionising radiation potentially harmful to humans?
12. Describe three safety precautions to be taken when handling radioactive materials. Explain why each precaution is necessary.

Supplement

13. Why is it safe to use an alpha-particle source in a smoke detector?
14. A tracer is to be used to find a blockage in an underground gas pipe buried 2 m down. The tracer is put into the pipe and the activity is monitored on the surface. Explain why the tracer should emit gamma radiation.

1 (a) Complete the table, which describes some key properties of the three particles that make up atoms. One line has been done for you.

Name	Relative charge	Location in the atom
proton	+1	nucleus
neutron		
electron		

[2]

(b) $^{12}_{6}C$ and $^{14}_{6}C$ are two isotopes of carbon.

(i) What are isotopes? [2]

(ii) The proton number for carbon is 6. Complete the table to show the number of each type of particle in neutral atoms of $^{12}_{6}C$ and $^{14}_{6}C$.

Isotope	Number of protons	Number of neutrons	Number of electrons
$^{12}_{6}C$			
$^{14}_{6}C$			

[3]

Supplement

(c) Describe how the scattering of alpha particles by thin metal foils provides evidence for the nuclear atom. [4]

2 (a) Compare the structure and properties of alpha radiation and gamma radiation.

You should include:

- what the radiations consist of

- ionising effect

- penetrating ability [6]

Supplement

(b) $^{192}_{77}Ir$ is an emitter of beta radiation. It can be used to treat tumours that measure a few centimetres across. A sample of iridium wire is implanted into the tumour where the abnormal cells are. Use the properties of beta radiation to explain why it is suitable for this application. [4]

3 $^{238}_{92}U$ is a radioactive isotope that decays by emitting alpha particles.

(a) What is radioactive decay? [2]

(b) Radioactive decay is a random process. What does *random* mean in this context? [2]

Supplement

(c) Complete the nuclear equation below which shows the decay of $^{238}_{92}U$:

$^{238}_{92}U = \ \overset{.........}{......}Th + ^{4}_{2}\alpha$ [2]

(d) Another isotope of uranium, $^{238}_{95}U$, is a fuel in nuclear fission reactions. Describe the process of nuclear fission. [3]

4 A teacher measured the activity of a radioactive sample every 10 minutes for 1 hour. The teacher also measured the background radiation and subtracted this value from the recorded values to give the corrected count rate shown in the table.

Time (minutes)	Corrected count rate (counts per second)
0	70
10	52
20	40
30	31
40	24
50	17
60	15

(a) State **two** safety precautions the teacher should take when using radioactive materials. [2]

(b) (i) What is background radiation? [1]

Supplement

(ii) Explain why the teacher subtracted the measurements for background radiation from the data she collected. [1]

(c) Use the data in the table to plot a suitable graph and use the graph to find the half-life of the sample. [4]

(d) Suggest, with a reason, whether this radioactive source would be suitable to use in a smoke alarm. [2]

Answers

1 General physics

Length and time

1. A rule (marked in cm or mm).
2. For short times, the reaction time of the user can make a significant difference.
3. Use a pile of paper (e.g. a ream of 500 sheets). Measure the thickness to the nearest mm using the rule and then divide the measurement by the number of sheets to find the thickness of one sheet.
4. Micrometer screw gauges will measure to 0.01 mm whereas a rule can only measure to the nearest 1 mm (although you can *estimate* a little less than that).

Motion

1. Stationary/not moving.
2. Constant speed.
3. The graph would be getting steeper/ the gradient would be increasing.
4. Constant speed/velocity.
5. Constant acceleration/deceleration.
6. It is given by the area under the graph.
7. A vector quantity includes a reference to direction, while a scalar quantity does not.
8. Scalar quantities – speed, mass, temperature. Vector quantities – velocity, force, momentum. (There are other examples of course!)

Mass and weight

1. Weight is the gravitational force on an object, mass measures the amount of material in an object (and the resistance of an object to changes in its motion).
2. Weight = mass × gravitational field strength.
3. Gravitational force causes an object to accelerate downwards. The faster it gets, the bigger the force of air resistance opposing the motion. When the two forces are equal the object falls at a constant speed – terminal velocity.

Density and pressure

1. Measure the dimensions of the object and calculate the volume.
2. Measure the volume of water displaced when the object is immersed.
3. Density = mass / volume.
4. Pressure = force applied / area.
5. An example would be using large tyres on a tractor, there are many other examples.

6. An example would be the point of a pin where the area is small, there are many other examples.
7. Pressure = depth × density × *g*.

Forces 1

1. Forces can change the shape, size or motion of an object.
2. The combined effect of two or more forces.
3. If the forces act in the same direction add the forces, if they act in opposite directions subtract the forces.
4. The object will remain at rest or, if it was already moving, continue in a straight line at a constant speed.
5. m/s^2
6. See the notes in the experiment box on page 14.
7. The object will accelerate – this includes speeding up, slowing down or changing direction.
8. Force is proportional to extension (provided limit of proportionality is not exceeded).

Forces 2

1. Moment = force × perpendicular distance from the pivot.
2. Increase the force, increase the distance of the force from the pivot.
3. If an object is balanced, sum of clockwise moments = sum of anticlockwise moments.
4. Resultant force = 0 and total moment = 0.
5. The centre of mass is the point where we consider all the mass to act.
6. See the notes in the experiment box on page 16.

Momentum

1. Momentum = mass × velocity.
2. A vector quantity has a magnitude and a direction.
3. Momentum = 600 kg × 75 m/s = 45 000 kg m/s.
4. In a closed system, total momentum is always conserved (remains constant).
5. The momentum before it fires is zero, so if the bullet has a forward momentum the gun must have a backwards momentum to keep the total momentum zero.

Energy, work and power

1. (a) Throwing a ball, (b) lighting a lamp, (c) boiling water over a fire, (d) sending radio signals (many alternatives possible).
2. Total energy is always conserved (the total remains the same).
3. Work is done when a force causes an object to move.

4. Work done = energy transferred.
5. Dissipated means that the energy becomes more 'spread out' among a larger number of particles.
6. Kinetic energy = ½ × mass × (velocity)^2 and gravitational p.e. = mass × *g* × change in height.
7. Work done = force × distance moved in the direction of the force.
8. Power = energy transferred / time taken for the transfer.

Energy resources

1. A source of energy that we can usefully use to provide energy for society.
2. Electricity.
3. Renewable energy resources can be replaced over reasonably short time periods, non-renewable resources cannot.
4. See table on page 23.
5. See table on page 23.
6. See table on page 23.

Exam-style practice questions

1. (a) (i) C   [1]
    (ii) Graph is steepest/biggest gradient.   [1]
   (b) Speed = distance / time = (2 × 600) / 600 = 2 m/s. [1 for doubling distance given, 1 for converting time to seconds, 1 for answer.]   [3]
   (c) (i) Speed has magnitude only [1], velocity has magnitude and direction [1].   [2]
    (ii) 0/zero (m/s).   [1]
    (iii) Returned to starting point, so overall displacement is zero.   [1]
2. (a) Low centre of mass [1], wide base [1].   [2]
   (b) (i) $W = m × g = 700 × 10$ = 7000 N. [1 for calculation, 1 for unit.]   [2]
    (ii) $P = F / A = 7000 / 0.5$ = 14 000 Pa. [1 for transferring F from part (i), 1 for answer.]   [2]
   (c) a = change in velocity / time = 46 – 0 / 4 = 11.5 m/s^2. [1 for change in velocity, 1 for answer, 1 for unit.]   [3]
3. (a) (i) Weight (accept gravity/ gravitational).   [1]
    (ii) $W = m × g = 70 × 10$ = 700 [1] N [1].   [2]
   (b) (i) Label X on the horizontal part of the graph (ideally the higher one, but either is correct).   [1]
    (ii) *Any four of:* Weight causes skydiver to accelerate downwards ; air resistance

Answers

increases as the velocity increases ; when weight = air resistance resultant force is zero ; skydiver falls at constant velocity. [4]

(c) (i) Y marked where line falls suddenly. [1]

(ii) *Any three of:* Opening parachute increases area ; which increases air resistance ; which gives resultant force upwards ; which slows the parachute down ; which reduces the air resistance ; until weight = air resistance again and skydiver falls at constant velocity. [3]

4. (a) Needs a <u>comparison</u> – LED bulb costs 8× more [1], but lasts 40× longer [1] so cost per hour is less for LED bulb. To run for 40 000 hours costs $20 with LED bulb but $100 with filament bulbs [1]. [3]

(b) (i) $E = P \times t = 12 \times 600 = 7200$ J. [1 for time conversion, 1 for answer.] [2]

(ii) $E = 0.8 \times 7200 = 5760$ J. [1 for 0.8×, 1 for answer, allow 5600 J if 7000 used from (i)] [2]

(iii) (Wasted as) heat. [1]

5. (a) *Any suitable two*, e.g. wind, solar etc. [2]

(b) *Any suitable two*, e.g. coal, oil etc. [2]

(c) (i) *Any suitable two*, e.g. rising cost of non-renewable sources (make it more economic) ; environmental concerns (from the public) ; future planning to take account of reduced fossil fuel availability etc. [2]

(ii) *Any suitable two*, e.g. unreliability of sources ; technology for non-renewable sources cheaper historically ; non-renewable sources 'more concentrated' energy stores. [2]

2 Thermal physics

Simple kinetic molecular model of matter

1. Solid, liquid, gas.
2. Small particles (e.g. smoke particles) are seen to move in a random path, suggesting they are being hit by smaller, invisible particles.
3. The molecules are able to 'slide' over each other, so they can fill any spaces in the container.
4. The molecules are held in place by strong forces.
5. The molecules with the greatest energy are the ones that leave, so the remaining molecules have a lower mean energy.
6. Temperature, surface area, presence of a draught.
7. If the temperature increases, the pressure increases (or vice versa).
8. The molecules have further to travel between collisions with the container walls if the volume increases, so collisions happen less often, giving a lower pressure.

Thermal properties and temperature

1. The volume increases.
2. The liquid in a thermometer expands when heated, allowing us to measure temperature (there are many other examples).
3. The molecules gain energy and vibrate more vigorously. This causes them to move apart very slightly, meaning that the material overall fills a larger volume.
4. The scale is marked at two fixed points and the space in between is divided.
5. The liquid expands at the same rate across the range of temperatures to be measured.
6. *Any three of:* Quick response ; can be small ; can measure high temperatures ; can be used in remote locations.
7. How much heat energy is required to change the temperature by 1 °C.
8. Refer to the box on page 32.
9. At a change of state, the energy supplied alters the molecule structure, not the average energy of the molecules.
10. At boiling, the mean energy of the molecules is high enough to change from liquid to gas. In evaporation, only the most energetic particles have sufficient energy.
11. Refer to the box on page 33.

Thermal processes

1. Heat is transferred quickly in a good thermal conductor and slowly in a bad thermal conductor.
2. Metals have delocalised electrons, which can also transfer energy.
3. As air is heated it expands, making it less dense. Less dense air floats upwards over cooler, more dense, air.
4. A convection current is a cycle where the hotter parts of a fluid rise and cooler parts sink.
5. Infra-red.
6. Dull, black surfaces are the best absorbers and emitters.
7. Both factors increase the rate of infra-red emission.
8. See the box on page 36.

Exam-style practice questions

1. (a)

	Fixed volume	Fills the container	Fixed shape	Takes the shape of the container
Liquids	✓			✓
Gases		✓		✓

[1 for each correct line.] [2]

(b) Particles held together by strong forces [1] holding the particles in fixed positions [1]. [2]

(c) (i) Particles have a range of energies [1], those with the most [1] energy are able to leave the surface [1]. [3]

(ii) <u>Mean</u> energy of the remaining particles is reduced [1] so liquid at a lower temperature than skin [1] so heat conducts from runner to sweat [1]. [3]

2. (a) Good conductor [1] so transfers heat quickly [1]. [2]

(b) Vibration passed on from particle to particle [1] or by free electrons (since it's a metal) [1]. [2]

(c) *Any three of:* Conduction from pan to water [1] which expands [1] becoming less dense [1] and rises [1] colliding with potato and conducting energy to it [1]. [3]

(d) Similarity – transition from liquid to gas [1] ; difference – boiling at a specific temperature, evaporation at a range of temperatures/in boiling mean energy is sufficient to change state, in evaporation only the highest-energy particles can leave [1]. [2]

3. (a) Temperature. [1]

(b) Particles in liquid gain kinetic energy [1] moving apart [1] so the liquid <u>expands</u> [1]. [3]

(c) Quicker response [1], less energy removed from the material being measured [1]. [2]

(d) Markings on the scale are evenly spaced. [1]

4. (a) Air over the land heated to higher temperature, so it expands [1] becoming less dense [1] and so rises [1] as a convection current [1] – ALLOW reverse argument for air falling over the sea. [4]

(b) Sand has a lower specific heat capacity [1] so temperature change is greater for the same energy transfer [1]. [2]

Answers

(c) Temperature change = 30 °C [1]
c = energy / (mass × temperature change) = 49 800 / (2 × 30) [1]
= 830 [1] J/kg °C [1].   [4]

3 Properties of waves, including light and sound

General wave properties

1. Waves transfer <u>energy</u> without transferring <u>matter</u>.
2. In longitudinal waves, the vibrations are parallel to the energy transfer. In transverse waves, the vibrations are perpendicular to the energy transfer.
3. Check against the diagram on page 40.
4. Refraction is the change of speed at a boundary, usually causing a change in direction of the wave.
5. Diffraction is the spreading out of a wave as it goes through a gap (or past an edge).
6. The diffraction effect is biggest when the gap is about the same size as the wavelength of the wave.
7. Wavespeed = frequency × wavelength.

Light

1. Refer to the diagram on page 43.
2. Angle of incidence = angle of reflection.
3. The image in a plane mirror is upright, laterally inverted, as far behind the mirror as the object is in front.
4. One where the rays of light *appear* to come from – it cannot be formed on a screen.
5. Refer to the diagram on page 44.
6. Refractive index measures how big a change of speed/direction occurs at a boundary.
7. Refractive index, $n = \sin i / \sin r$ = speed in air / speed in material.
8. Total internal reflection is where a ray of light is reflected back into a material at a boundary instead of emerging.
9. Total internal reflection happens at an optically more dense to less dense boundary when the angle of incidence is bigger than the critical angle.
10. Refer to the diagram on page 46.
11. Refer to the diagram on page 46.
12. Refer to the diagram on page 47.
13. Dispersion is where white light splits into the separate colours of the visible spectrum.

Electromagnetic spectrum

1. Radio, microwaves, infra-red, visible, ultraviolet, X-rays, gamma.

2. The order would be reversed.
3. 3.0×10^8 m/s.
4. Refer to the table on page 49.
5. Microwaves cause water to heat up, so they can harm living tissue (which contains lots of water).
6. X-rays and gamma radiation damage living cells and can cause mutations, leading to cancers.

Sound

1. All sounds are caused by <u>vibrations</u>.
2. Sound is a pressure wave with energy transferred from particle to particle – there are no particles in a vacuum.
3. Compressions are regions where the particles are closer together than usual and rarefactions are regions where the particles are further apart than usual as the wave passes through.
4. Refer to the box on page 50.
5. The particles are closer together in a solid, so the energy is transferred more quickly.
6. High pitch = high frequency, loud = large amplitude.
7. 20 Hz to 20 000 Hz.
8. Sound with frequencies above the range of human hearing.

Exam-style practice questions

1. (a) D   [1]
   (b) B   [1]
   (c) Vibration is perpendicular [1] to the direction of energy transfer [1].   [2]
   (d) (i) Waves spreading out from the gap [1] with wavelength maintained as before the gap [1].   [2]
   (ii) Diffraction occurs most when the size of the gap approximately matches the wavelength of the waves.   [1]
2. (a) (i) Normal line drawn at right angles to the mirror at the point where the incident ray strikes [1] with reflected ray leaving the mirror at the same point at the correct angle of reflection (judged by eye) [1].   [2]
   (ii) Angle of incidence marked between incident ray and normal line [1]; angle of reflection marked between reflected ray and normal line [1].   [2]
   (iii) Equal to.   [1]
   (iv) Cannot be formed on a screen/point where rays appear to come from.   [1]

(b) (i) Change of speed/direction [1] when a wave crosses a boundary [1].   [2]
   (ii) Speed in glass = 3×10^8 / 1.5 [1] = 2×10^8 m/s [1].   [2]
(c) (i) Going from more optically dense to less optically dense [1] at greater than the critical angle [1].   [2]
   (ii) Can direct light along a non-straight path.   [1]
3. (a) (i) Microwaves [1] ; infra-red [1].   [2]
   (ii) Ultraviolet.   [1]
   (b) Increasing frequency.   [1]
   (c) (i) 3×10^8 m/s.   [1]
   (ii) Frequency = speed / wavelength = 3×10^8 / 1500 [1] = 200 000 [1] Hz / hertz [1].   [3]
4. (a) Vibration is parallel [1] to the direction of energy transfer [1]. [2]
   (b) Speed = distance / time = (2 × 135) [1] / 0.8 [1] = 337.5 (m/s) [1].   [3]
   (c) Particles are closer together [1] so energy transfer is quicker [1].   [2]
   (d) (i) Ultrasound is high-frequency sound [1] above 20 000 Hz/ range of human hearing [1].   [2]
   (ii) Ultrasound is non-ionising [1] so safer [1].   [2]

4 Electricity and magnetism

Simple phenomena of magnetism

1. A magnet has its own magnetic field, a magnetic material can be affected by a nearby magnetic field.
2. From N-poles to S-poles.
3. They show the direction of the force on a N-pole – that can only be in a single direction.
4. Sprinkle iron filings in the area or note the direction of a plotting compass needle.
5. It does not stay permanently magnetised.
6. Stroke in one direction with a magnet ; place in a coil carrying direct current ; hammer gently in a magnetic field.
7. Wrap a coil of wire around an iron core and pass a current through the coil.
8. Circular around the wire, stronger nearer the wire.
9. The current magnetises the coil, which attracts the contacts together, switching on a second circuit.
10. The relative strength of the field at that point.

Answers

Electrical quantities

1. Conductors allow electric charge to pass through easily (due to free electrons), while insulators do not.
2. Insulators.
3. Attract small pieces of paper ; deflect a thin stream of water from a tap.
4. Bring another, charged, object close by.
5. An ammeter.
6. Digital meter displays numbers directly, analogue meter usually has a needle moving across a scale.
7. Current = charge / time.
8. Both measured in volts as they measure energy transferred per unit charge. E.m.f. relates to energy transferred into electrical energy at the supply, potential difference relates to energy transferred to other forms across a component.
9. Transferring energy from a source to circuit components and then to the surroundings.
10. Power = voltage × current.
11. Resistance = voltage / current.
12. Refer to the graphs on page 61. The resistor has a constant resistance so produces a straight line. The filament lamp resistance increases as its temperature increases, so the line is curved.
13. Increasing length increases the resistance (proportional). Increasing the diameter decreases the resistance (inversely proportional to cross-sectional area).

Electric circuits

1. Refer to the diagram on page 63.
2. There is only one pathway, so all the electrons travel the same way and current is related to the number of electrons following a pathway.
3. The electrons can take alternative pathways at different parts of the circuit, so there will not necessarily be the same number on each pathway.
4. Add the values together.
5. As the temperature increases, the resistance of the thermistor decreases.
6. Diodes only allow current / electrons to flow in one direction.
7. A circuit with two resistors in series, used to split the supply voltage into two.
8. A circuit using a diode that converts alternating current (a.c.) to direct current (d.c.).

Digital electronics

1. Digital signals have 'high' and 'low' values only, analogue signals can have any value.

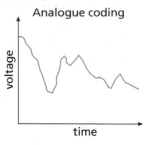

Analogue coding

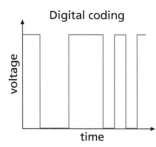

Digital coding

2. A circuit designed to give a specific output signal, depending on its input signals.
3. A table summarising the operation of a logic gate or a circuit of connected logic gates.
4. Refer to the tables on page 66.

Dangers of electricity

1. Damaged insulation or cables ; overheating cables ; damp conditions.
2. If the current gets too high the fuse melts, switching off the circuit, before any other parts are damaged/ overheat/catch fire.
3. If the 13 A fuse is used it will not melt at currents up to 13 A. These are likely to damage an appliance designed to run at 2.3 A (or overheat the cables).
4. It responds more quickly and can be reset (after the fault is corrected).
5. Plastic is an insulator, so the user cannot get a shock.
6. If the case becomes live, the earth wire provides a low-resistance path to the ground, allowing a large current to flow, which will melt the fuse and switch off the circuit.

Electromagnetic effects

1. The conductor should be moving in a magnetic field or be in a moving magnetic field.
2. Increase the speed of the motion ; increase the strength of the magnetic field ; wrap the conductor as a coil.
3. In d.c. the electrons always travel in the same direction, in a.c. the electrons reverse their direction regularly.
4. Refer to the graph on page 71.
5. They are the connections at the end of the coil. They brush against the external circuit, allowing a current to flow.
6. Change voltages.
7. In a step-up transformer, the secondary voltage (the output voltage) is larger than the primary voltage (the input voltage). In a step-down transformer the reverse is true.
8. Less energy is wasted in the cables.
9. The conductor experiences a force. If the current and magnetic field are at right angles to each other, the force will be at right angles to both.
10. Reverse the direction of the current ; reverse the polarity of the magnetic field.
11. Refer to the diagram on page 72.
12. The two sides of the coil experience forces in opposite directions, creating a turning moment around the axle.
13. Reverse the direction of the current ; reverse the polarity of the magnetic field.

Exam-style practice questions

1. (a) Field pattern approximately correct shape (compare with main text) [1] with at least one arrow pointing from N to S [1] and no field lines crossing [1]. [3]
   (b) Place compass in the field, mark positions of ends of compass needle [1] ; move compass so end of the needle is at one of the points marked [1] ; repeat the process and field line is found by joining the marks together [1]. [3]
   (c) Non-magnets also attracted [1] through induced magnetism [1]. [2]
   (d) (i) *Any two of:* Current [1] ; number of turns on coil (per metre) [1] ; presence of iron core [1]. [2]
       (ii) Magnetism can be switched off (when current is turned off) ; strengthens magnetic field. [1]
2. (a) Circuit diagram to include source, ammeter and wire in series [1] with voltmeter in parallel across the wire [1] ; measure current and voltage [1] for a range of lengths [1] and diameters [1] ; calculate resistance using $R = V / I$ [1]. [6]
   (b) As the area increases the resistance decreases [1] such that double the area gives half the resistance [1]. [2]

Answers

(c) To stop the temperature of the wire changing [1] which would also change the resistance [1]. [2]

3. **(a)** Can be controlled separately [1] ; if one fails the other still works [1]. [2]
   (b) Current = power / voltage = 1200 / 230 [1] = 5.2 (A) [1]. [2]
   (c) (i) 7 A [1]
   (ii) Needs to be the next highest above working current. [1]
   (d) If the (metal) casing becomes live [1] the earth wire provides a low-resistance path to the ground [1] so a large current flows [1] melting the fuse (and switching off the circuit) [1]. [4]
4. **(a)** Increase voltages. [1]
   (b) Transmits at low current [1] which reduces heating due to resistance [1] so less energy is wasted [1]. [3]
   (c) (i) Primary coil labelled (input voltage coil) [1] ; iron core labelled [1]. [2]
   (ii) 100 / 400 = 12 / output voltage [1] so output voltage = 12 × (400 / 100) [1] = 48 (V) [1]. [3]

5 Atomic physics
The nuclear atom
1. The nucleus has a positive charge.
2. Electrons have a negative charge.
3. Large-angle deflection was only possible if the matter in the atom was very dense. The overall mass was known so this mass must be concentrated in a small region – the nucleus.
4. Refer to table on page 77.
5. X is the chemical symbol, A is the nucleon number (number of protons + number of neutrons), Z is the proton number (number of protons).
6. Isotopes are atoms with the same number of protons but different numbers of neutrons.
7. The splitting of (large) nuclei into two smaller nuclei following the absorption of a neutron, releasing energy.
8. Each fission releases additional neutrons which can then strike other nuclei causing fission to continue.
9. Nuclear fusion is the joining of smaller nuclei to form a larger nucleus, releasing energy.

Radioactivity
1. The nucleus will emit particles or waves to achieve a more permanent state.

2. It cannot be predicted exactly when a nucleus will decay, or which one will decay.
3. The emissions can remove electrons from atoms in their path by collision.
4. Background radiation is radiation that is always present around us, from sources such as rocks or space.
5. A Geiger–Muller tube or a photographic film.
6. Refer to the table on page 79.
7. Alpha particles are relatively massive so it is highly likely they will collide with atoms in their path, knocking out electrons as they lose energy and slow down.
8. The number of protons in the nucleus changes.
9. The half-life is the time it takes on average for half the nuclei present to decay.
10. Find the time at which the sample has a particular activity and then the time for half that activity. The half-life is the difference between these two times.
11. Ionising radiation can damage cells, causing mutations in the cells, or can destroy cells.
12. Refer to the list on page 81.
13. The range and penetrating power of alpha particles are so low that they present no hazard outside the casing of the smoke alarm.
14. Only gamma radiation would be able to penetrate through the 2 m of earth to the surface.

Exam-style practice questions
1. **(a)** [1 for each correct column.]

Name	Relative charge	Location in the atom
neutron	0	nucleus
electron	−1	orbiting/ outside nucleus

[2]

   (b) (i) Atoms with the same number of protons [1] but different numbers of neutrons [1]. [2]
   (ii) [1 for each correct column.]

Isotope	Number of protons	Number of neutrons	Number of electrons
$^{12}_{6}$C	6	6	6
$^{14}_{6}$C	6	8	6

[3]

   (c) Most alpha particles undeflected [1] so most of atom is empty space [1] ; a few alpha particles deflected through large angles [1] indicating the presence of dense, central nucleus [1]. [4]

2. **(a)** [1 for each relevant point (organising in a table is a good idea).]

	Alpha radiation	Gamma radiation
Consists of/ nature	Two protons and two neutrons	Electro-magnetic radiation
Ionising effect	High	Low
Penetrating ability	Low/paper/ few cm of air	High/thick concrete/ thick lead

[6]

   (b) Penetrates a few cm in flesh [1] so will not harm cells outside the tumour [1] ; has the ability to cause ionisation in cells [1] so can destroy abnormal cells [1]. [4]
3. **(a)** The emission of particles/radiation [1] from an unstable nucleus [1]. [2]
   (b) Cannot predict *when* a particular atom will decay [1] or *which* particular atom will decay [1]. [2]
   (c) Top number 234 [1], bottom number 90 [1]. [2]
   (d) Nucleus absorbs a neutron [1], splits into two smaller nuclei [1], releasing further neutrons (plus energy) [1]. [3]
4. **(a)** *Any two of:* Handle with tongs ; limit exposure time ; keep students a safe distance away ; only have one source out at a time ; wear gloves ; any suitable relating to use in a classroom. [2]
   (b) (i) Radiation present all the time from the environment. [1]
   (ii) So that the readings used relate only to the source being investigated. [1]
   (c) Graph of count rate against time – correct scales and plotting [1], suitable line [1], at least two values read from graph for count rate and then half of that value [1], correct half-life value (from plotted line) [1]. [4]
   (d) Not suitable (no mark) – the half-life is too short [1] so it would need replacing too regularly to be practicable [1]. [2]

Glossary

Acceleration – the rate of change of velocity. Calculated using $a = v / t$ and measured in m/s^2.

Air resistance – a resistive force on an object caused by friction with the air. Air resistance increases as the speed of the object increases.

Alpha radiation – alpha radiation is highly ionising, has a short range in air, and consists of two protons and two neutrons, giving an alpha particle a relative charge of +2.

Alternating current (a.c.) – electric current where the electrons change the direction of their movement periodically.

Amplitude – the maximum displacement of a wave from the equilibrium position, measured in m.

Analogue signal – a signal (voltage) that can have any value.

Atom – smallest particle of a chemical element. Consists of protons, neutrons and electrons. The number of protons is equal to the number of electrons in a neutral atom.

Background radiation – the radiation that is present in the environment, caused by natural sources (such as rocks or space) or artificial sources (such as medical sources).

Beta radiation – beta radiation is moderately ionising, is blocked by thin metal sheets, and consists of an electron, giving a beta particle a relative charge of –1.

Boiling point – this is where the average energy of particles in a liquid is enough to leave the liquid and enter the gas state. Boiling happens at a specific temperature.

Brownian motion – the random motion of small particles which provides evidence for the molecular model.

Calibration – the process of correctly placing the scale on a measuring instrument.

Characteristic curve – a graph of current against voltage for a component, such as a filament lamp or a diode.

Circuit breaker – a switch that performs the same job as the fuse. Switches more quickly than a fuse and can be reset once the fault in the circuit is corrected.

Compression – region in a longitudinal wave where the particles are a little closer together than their equilibrium positions.

Conduction – the process of transferring heat energy from one particle to the next by collision. Most important in solid materials.

Convection – the process of transferring heat energy in fluids where higher temperature regions expand and float upwards, forming a convection current.

Converging lens – a lens with a convex shape that refracts rays parallel to the principal axis towards the principal axis.

Critical angle – the angle of incidence at a more optically dense to less optically dense boundary when the refracted ray emerges along the boundary (i.e. the angle of refraction is 90°).

Density – a measure of how compact a material is. Calculated using $\rho = m / V$ and measured in kg/m^3.

Diffraction – the spreading out of waves as they move through a gap or around an edge.

Digital signal – a signal (voltage) that is either high or low (often noted as 1 or 0).

Direct current (d.c.) – electric current where the net movement of electrons is always in the same direction.

Dispersion – the separating of white light into separate colours when it passes through a triangular prism.

Distance–time graph – graph showing how far an object travels over time. The gradient of the graph gives the speed.

Earth wire – a safety device that provides a low-resistance path to the ground if the metal casing of an appliance becomes live.

Efficiency – a measure of how much energy is usefully transferred. Calculated from efficiency = useful energy output / total energy input.

Electric charge – a measure of how much an object is affected by electric forces, measured in C.

Electric current – a net movement of electrically charged particles. In circuits this is a net movement of electrons. Calculated using $I = Q / t$ and measured in A.

Electric field – a region where a charged particle experiences a force. The direction of the electric field at any point is the direction of the force on a positive charge at that point.

Electric power – a measure of the energy transferred per second in a circuit. Calculated using $P = I \times V$ and measured in W.

Electrical conductor – a material in which some electrons are free to move within the structure.

Electrical insulator – a material in which all electrons are bound within atoms, with none available to pass along the material.

Electrical working – a measure of the energy transferred in a circuit. Calculated using $E = V \times I \times t$ and measured in J.

Electromagnet – a magnet made by passing an electric current through a coil of wire, usually wrapped around an iron core.

Electromagnetic induction – the process of generating an e.m.f. across a conductor when there is relative motion between the conductor and a magnetic field. The polarity of the e.m.f. is given by Fleming's right-hand rule.

Electromagnetic spectrum – a family of waves including radio waves, microwaves, infra-red, visible light, ultraviolet, X-rays and gamma radiation. All travel at the same speed in a vacuum.

Electromotive force – a measure of the energy transferred per unit charge to electrical energy in a power supply. Calculated using $E = W / Q$ and measured in V.

Electron – particle with a negative charge. Electrons orbit outside the nucleus of an atom. Their motion is responsible for electrostatic effects and for currents in circuits.

Electrostatic charge – materials gain an electrostatic charge when they have a shortage or excess of electrons, leaving the material with an overall positive or negative charge.

Energy – a measure of the work that an object is able to do. Measured in J.

Energy resource – a store of energy that can be used to provide useful energy for society.

Equilibrium – an object is in equilibrium if the resultant force on it is zero and the resultant moment on it is zero.

Evaporation – this is where some particles in a liquid have enough energy to leave and enter the gas state. Evaporation can happen at any temperature.

Expansion – the increase in the volume of a material when the temperature is increased, due to the increased motion of the particles.

Focal length – the distance from the centre of a lens to the principal focus.

Frequency – the number of waves transferred per second, measured in Hz.

Fuse – a circuit component designed to be the first part of the circuit that melts (switching off the circuit) if the current gets too high and the circuit overheats.

Gamma radiation – gamma radiation is weakly ionising, is blocked by thick lead or thick concrete, and consists of high-frequency electromagnetic radiation. Gamma radiation is uncharged.

Half-life – the time it takes for the activity of a radioactive source to reduce to half its initial value. Alternatively, half-life can be described as the time it takes for half the nuclei in a sample to decay.

Half-wave rectification – converting a.c. to d.c. using a diode.

Hooke's law – the extension of a spring is proportional to the force stretching it, providing the limit of proportionality is not exceeded.

Induced charge – where a material gains a temporary charge in an electric field.

Induced magnetism – where a magnetic material becomes a temporary magnet in a magnetic field.

Ionising radiation – emissions from nuclei that can ionise (knock electrons off) atoms in their path.

Isotopes – nuclei that have the same number of protons (and so are atoms of the same elements), but with different numbers of neutrons.

Glossary

Kinetic energy – energy due to motion. Calculated using $E = \frac{1}{2} \times m \times v^2$ and measured in J.

Logic gate – a digital circuit where the output (high or low) depends on the combination of input voltages.

Longitudinal – wave motion where the vibrations are parallel to the direction of energy transfer.

Magnetic field – the region around a magnet where it can affect magnetic materials.

Magnetic pole – the ends of a magnet where the magnetism is strongest.

Mass – a measure of the amount of matter in an object, measured in kg.

Micrometer screw gauge – instrument used to measure small lengths. Typically, micrometer screw gauges measure to the nearest 0.01 mm.

Molecular model – the idea that all matter is made of tiny particles.

Moment of a force – the turning effect of a force. Calculated using moment = $F \times$ perpendicular distance from the pivot and measured in Nm.

Momentum – a measure of the motion of an object. Calculated using momentum = $m \times v$ and measured in kg m/s.

Motor effect – the process where a force is exerted on a current-carrying conductor in a magnetic field. The direction of the force is given by Fleming's left-hand rule.

Normal line – a construction line drawn at 90° to a surface at the point where the waves arrive.

Nuclear fission – the process where a large nucleus absorbs a neutron and splits into two, smaller nuclei with the release of energy and additional neutrons.

Nuclear fusion – the process where two small nuclei join together to make a larger nucleus, releasing energy.

Nucleon number – the number of nucleons (protons and neutrons) in a nucleus. Also called the mass number.

Nucleus – the central part of an atom, consisting of positively charged protons and neutrons which have no charge.

Nuclide – a particular type of nucleus.

Parallel circuit – circuit in which components are connected together side by side.

Pitch – how high or low a sound is; higher pitch relates to higher frequency.

Potential difference – a measure of the electrical energy transferred per unit charge to other forms between two points in a circuit. Calculated using $V = W / Q$ and measured in V.

Potential divider – a circuit consisting of two (or more) resistors connected in series, used to split the supply voltage into fractions.

Potential energy – energy that is stored, measured in J. Gravitational potential energy is calculated using $E = m \times g \times h$.

Power – power measures the rate of energy transfer. Calculated using $P = E / t$ and measured in W.

Pressure – a measure of how 'spread out' a force is over an area. Calculated using $P = F / A$ and measured in Pa. The pressure at a depth h in a liquid of density ρ is calculated using $p = h \times \rho \times g$.

Principle of moments – if an object is balanced, then the sum of the clockwise moments is equal to the sum of the anticlockwise moments.

Proton number – the number of protons in a nucleus. Also called the atomic number.

Radiation – the process of transferring heat energy as electromagnetic infra-red waves.

Radioactivity – the emission of particles or energy from unstable nuclei in atoms.

Random process – radioactive emission is a random process, which means that the behaviour of individual nuclei cannot be predicted exactly.

Rarefaction – region in a longitudinal wave where the particles are a little more spread out than their equilibrium positions.

Real image – an image formed where the rays of light actually meet – if a screen is placed at that point an image would be seen.

Reflection – the bouncing of waves from a barrier. The angle of incidence is equal to the angle of reflection.

Glossary

Refraction – the change in speed of waves when moving from one medium to another. Refraction causes a change in direction of the waves at the boundary unless the waves arrive perpendicular to the boundary.

Refractive index – a measure of how much the speed and direction of a wave is changed at a boundary. Calculated using $n = \sin i / \sin r$.

Relay – an electromagnetic switch in which a coil becomes magnetised, attracting two contacts together which then complete a second circuit.

Resistance – a measure of how difficult it is for electrons to transfer along a conductor. Calculated using $R = V / I$ and measured in Ω.

Resultant force – the overall effect of a number of forces. If the resultant force is not zero, then an object will accelerate, calculated using $F = m \times a$.

Scalar quantity – a quantity that includes a magnitude (along with its unit) only. Examples include a mass of 20 kg or an energy of 30 J.

Series circuit – circuit in which components are connected together in line, one after another.

Specific heat capacity – the energy required to change the temperature of 1 kg of a material by 1 °C.

Specific latent heat – the energy transferred when 1 kg of a material changes state. The specific latent heat of fusion refers to the solid/liquid transition and the specific latent heat of vaporisation refers to the liquid/gas transition.

Speed – a measure of how quickly an object is moving. Calculated using $v = s / t$ and measured in m/s.

Speed–time graph – graph showing the speed of an object as it moves. The gradient of the graph gives the acceleration. The area under the graph gives the distance travelled.

Terminal velocity – a constant velocity reached by objects where the forces are balanced. For example, the speed reached by a parachutist when their weight is balanced by the air resistance.

Thermal capacity – the energy required to change the temperature of an object by 1 °C.

Total internal reflection – waves are reflected at a boundary when (1) the boundary is from less optically dense to more optically dense, and (2) the angle of incidence is greater than the critical angle.

Transformer – a device designed to increase or decrease voltages. Consists of two coils wrapped around a soft iron core.

Transverse – wave motion where the vibrations are perpendicular to the direction of energy transfer.

Truth table – a table that summarises the inputs and outputs of a logic gate or a combination of logic gates.

Ultrasound – sound with a frequency above the range of human hearing (i.e. above 20 000 Hz).

Vector quantity – a quantity that includes a magnitude (along with its unit) and a direction. Examples include a velocity of 2 m/s North or a force of 20 N down.

Virtual image – an image that the rays of light appear to come from, the image cannot be formed on a screen placed at that point.

Wavelength – the distance from one peak in a series of waves to the next, measured in m.

Wavespeed – a measure of how quickly a wave is travelling. Calculated using $v = f \times \lambda$ and measured in m/s.

Weight – the gravitational force on an object. Calculated using $W = m \times g$ and measured in N.

Work – work is done when energy is transferred. Calculated using $W = F \times d$ and measured in J.

Notes

Notes